Newton on Matter and Activity

Newton on Matter and Activity

ERNAN McMULLIN

University of
Notre Dame Press: NOTRE DAME ~ LONDON

Library of Congress Cataloging in Publication Data

McMullin, Ernan, 1924–
    Newton on matter and activity.

    Includes bibliographical references.
    1. Matter—History—Sources. 2. Motion—History—
Sources. 3. Newton, Isaac, Sir, 1642–1727. I. Title.
QC171.2.M3   531   77-82480
ISBN 0-268-01342-X

Manufactured in the United States of America

Contents

Acknowledgments

THIS WORK WAS substantially completed during a period of research support from the National Science Foundation at Cambridge University in 1973–74. I am indebted to the Foundation for its continuing support.

I wish to express my appreciation to D. T. Whiteside for his helpful comments on a first draft of the work, completed in May 1969. My special thanks go to J. E. McGuire, whose articles on Newton were an indispensable resource for me and whose detailed comments on successive drafts of the work were of enormous help. Discussions with Alan Shapiro and Cecil Mast aided me through some tangled stretches.

I am grateful to the University Library in Cambridge for permission to work with the Newton manuscript material, to the staff of the Research Library at the Whipple Science Museum, Cambridge, for their generous provision of research resources, and to the Fellows of St. Edmund's House for their hospitality during my stay in Cambridge.

One point of editorial usage may be noted here. Typography has been pressed into the service of clarity in a way which is happily becoming standard in philosophical works. Single quotes are used for *mention* only, that is, in order to name the expression they enclose. Double quotes are used not only for quoting material but also (in the case of words or short phrases) to indicate that the expression that they enclose is being used in some special sense. Italics are used for emphasis, or to indicate that the expression italicized is borrowed from a language other than the main one of the text. Finally, punctuation does not appear within quotes, unless it is part of the material being quoted.

Ernan McMullin

Introduction

IN THE STORY of the concept of matter, Newton plays a
paradoxical role. On the one hand, he struggled with the
intricacies of this concept for sixty years while building his
system of the world around it. Yet on the other hand, he
provided scientists with a neat and manageable substitute for
it, one which would later supplant the older concept in the
explicit symbolic systems of modern science. If one defines
Newton's achievement in mechanics by the text of the *Principia*,
an assumption that philosophers of science have often
found convenient, it would be plausible to say that 'matter'
occurs there only in the phrase 'quantity of matter', and thus
that in Newton's own work the function of the concept of
matter has already been taken over by that of mass. Yet, as we
shall see, nothing could be further from the truth. In the
immense effort to construct an adequate natural philosophy
around and beyond the *Principia*, an effort which occupied
Newton through much of his later life, the concept of matter
played an altogether crucial role.

When Newton said in the *Principia* that he intended only to
give a "mathematical notion" of the forces he discussed,
"without considering their physical causes and seats", this was
not at all the prescient positivism that historians have sometimes made it.[1] Newton not only thought that these forces *had*
"physical causes and seats" but spent much of the rest of his
scientific career trying to sort them out. In the *Principia* itself,

1

he gives a clue to what he felt to be the proper order in this matter:

> In mathematics we are to investigate the quantities of forces with their proportions consequent upon the conditions supposed; then, when we enter into physics, we compare those proportions with the phenomena of Nature that we may know what conditions of those forces answer to the several kinds of attractive bodies. And this preparation being made, we argue more safely concerning the physical species, causes and proportions of the forces.[2]

There are thus *three* stages here: mathematics (mathematical analysis of the consequences of different laws of force), physics (establishing which laws of force actually obtain in Nature), and finally the more "philosophical" part of the task of the natural philosopher, the discussion of such issues as the causes of various forces. Newton deliberately restricted the *Principia* to the first two stages, which he felt had to be carried out first, before the third stage could profitably be entered upon.

Besides the question of the nature of force, another important problem left unresolved in the *Principia* was that of the distinction between body and void, around which so much of the formal discussion of motion in the book hinged. Both of these problems reduced to that of the proper understanding of the concept of matter. If matter is to be identified as the inert and passive principle of motion, as Newton, following the neo-Platonic tradition going back through More, tended to make it, then there must be something else ("active principles") to explain how change originates. But where are these "active principles" to be located? In the vortices of a Cartesian "aether"? This would only push the question one stage further back. Besides, an aether theory was incompatible (as the *Principia* showed) with the lack of resistance to the planets in their motions around the sun. On the other hand, if one were to postulate atoms moving in a void, how is the transmission of action—especially electrical and magnetic action—to be explained? Further, the Cartesians had seemed to make of the universe an all-sufficient machine. Newton opposed this, on theological grounds, and for a time at least preferred to invoke (as Malebranche had done before him) God's action as the immediate source of change of motion, as the single "active principle" needed.

These were issues Newton could not avoid facing. It was all very well for the special purposes of the *Principia* to confine himself to "mathematics" and "physics". But as he knew perfectly well, many ambiguities and inconsistencies still surrounded the central issue: how is action transmitted? The theorems of the *Principia* did not really help one to understand how gravitational motion occurred, what its causes were. In that sense, they did not *explain* it.[3] They simply described it, and cancelled out the force-expressions by substituting specific laws of force in the calculation of orbits. There was still much hard philosophical analysis to be done if the nature of force and the transmission of action were to be properly understood, and the assumptions of the *Principia* adequately protected against attack. And so, in the early 1690's, Newton set about making drafts for a revision of Book III of the *Principia*, a revision which was never completed. Work on the *Opticks* also progressed, and the first edition appeared in 1704, to be followed shortly after by an enlarged Latin edition (*Optice*, 1706). The most striking feature of the *Opticks* was the lengthy set of "Queries" appended to it, dealing with the interaction of matter and light.[4] This was a topic on which Newton had very definite views, but since he could not "deduce" his corpuscular model from the phenomena, he preferred to propose his hypotheses as speculative queries in order to encourage "a farther search to be made by others".[5] Their avowedly tentative form marks them off from the rest of Newton's published work, and makes them the most significant source, perhaps, for the most general categories of matter and action that informed his researches.

The correspondence with Cotes preparatory to the second edition of the *Principia* (1713)[6] shows the extent to which the Newtonian group was preoccupied with the criticisms leveled against them by Cartesians concerning the transmission of action and the nature of force. The disagreement finally broke into the open with the famous exchange of letters between Leibniz and the Newtonian, Clarke, on a series of issues centered around the nature of space and the mode of God's action in it.[7] Though now in his seventies, Newton continued to make corrections and to draft constant revisions of his mechanical concepts in preparation for the third edition of the *Opticks* (1717) and the third edition of the *Principia* (1726). But much of what he wrote in the forty years between the first and third editions of the *Principia* concerning the theory of mechanical action underlying his great work remained unpublished until scholars in the last couple of decades began to sift the enormous mass of manuscript material he left behind.[8] It has seemed worth stressing this because these hitherto unpublished materials are probably more relevant to uncovering the gradual transformation worked by Newton in basic concepts such as *matter* than are his published works, where most signs of the underlying (and still speculative) categorial transformations have been carefully tidied away.

CHAPTER ONE

The Universal Qualities of Matter

AFTER THE *Principia* appeared, it was evident that some clarification was needed of the term 'matter' or 'body' (usually taken to be synonymous) if any headway were to be made in the growing controversy about the reality of the void which the mechanics of the *Principia* seemed to postulate. How could one decide whether the void was "corporeal" or "incorporeal", or whether the subtle media being spoken of in connection with gravitation were "material" or not, unless one knew what the identifying properties of matter were, the properties that presumably marked off "corporeal" from "incorporeal" entities? Hobbes had said that only material things existed, so that such terms as 'immaterial' were dismissed by him as meaningless; he did not, therefore, have to worry about the *differentiae* of matter. But Newton (like most of his contemporaries) still used such phrases as "any medium whatever, whether corporeal or incorporeal",[9] and they obviously had significance for him.

§1.1 The Limits of Transmutability

A more pressing reason, however, for Newton's search for the identifying properties of matter appears to have been a logical difficulty that developed when he began to work out his views concerning the unity of matter. In *Principia* I, he had asserted that "any body can be transformed into another body

5

of any kind whatsoever, and can assume successively all intermediate degrees of quality".[10] No evidence was given for this far-reaching "Hypothesis" (as he calls it there), and at first sight it is hard to see what function it has in the overall structure of the work. Did it derive from his early interest in alchemy, where universal transmutability had always been assumed without question?[11] Or from a tacit postulate of the basic continuity of all forms of change, required if the mathematical analysis of the *Principia* were to apply to it? Or perhaps from his youthful speculations about the aether, when he had surmised that "perhaps all things may be originated from aether" wrought "by condensation into various forms"?[12] But now he had a far more powerful explanatory model than aether condensations. His new concept of force, allied to a plausible sort of atomism in which the primordial particles were all of the same density and kind,[13] could in principle provide an unlimited transmutability of *all* bodies, the constituent particles themselves alone excepted.[14] In a draft-conclusion for *Principia* I, he speculates that the differences between natural kinds may be explained in terms of "particles coalescing in new ways... by means of attractive forces", which allows him to conclude that "the matter of all things is one and the same, which is transmuted into countless forms by the operations of nature".[15] Some years later, in Query 31(23), he repeated this view: "The changes of corporeal things are to be placed only in the various separations and new associations and motions of these permanent particles".[16]

In his own eyes, this was no new idea. In some notes made in the early 1690s for a revision of the first edition, he transforms the "Hypothesis" into an "Axiom" (the word 'hypothesis' was not in favor with him by this time!) and adds this explanatory note to it:

This is laid down by almost all philosophers. All teach that from some common matter taking on various shapes and textures all things arise and are dissolved back into the same matter by the privation of the shapes and textures: like the Cartesians who want matter to be extension divided and modified in various ways; the Peripatetics who imagine some kind of primary matter which is formless but capable of all forms... Pythagoras, Democritus and the rest who compose things either from atoms combined in various ways or from the four elements which arise out of a common matter. [17]

This is a significant passage. It shows Newton not only conscious of the varieties of the "matter tradition" but anxious at this stage to present himself as its inheritor. Matter is the substratum of change and consequently that which underlies differences of kind. In *Principia* II, he lumped Aristotelians and Cartesians together, implicitly identifying "form" with shape, when rejecting the view that "form" could of itself affect the weight of a body (and by implication, the suggestion that change of form of the constituent particles could of itself alter the quantity of matter). [18]

Newton gradually came to believe that he would have to limit his original transformation hypothesis in order that *mechanical* properties would remain invariant. After all, if solidity could be "remitted" (decreased) at all, it was conceivable that it could be "taken away" entirely, [19] yet this must clearly be excluded, since it would entail that a body could cease to be subject to mechanics, that is, could cease to be a body. And if inertia were taken to be a quality, it could not be treated as variable in a given body without violating the basic laws of the *Principia*. So his Hypothesis was gradually transformed (as his notes show) until it evolved into something quite different, the third of the "Rules of Reasoning" in the second edition:

The qualities of bodies which admit neither increase nor decrease of intensity, and which are found to belong to all bodies within reach of our experiments are to be esteemed the universal qualities of all bodies whatsoever. [20]

And he lists these qualities as "extension, hardness, impenetrability, mobility and inertia". They are to be regarded as the "universal" qualities of matter, qualities an entity would *have* to possess in order to be regarded as material, by contrast with such qualities as heat, wetness, light, color, transparency, and acidity, which, in an earlier draft,[21] he lists as examples of qualities which admit of change in intensity and are therefore, it appears, disqualified from "universal" status.

Can they be called "essential" qualities also? Commentators on this passage have often assumed that they can,[22] though Newton professed himself opposed to this usage. There were two reasons for his uneasiness about the term 'essential' in this context. One was a general skepticism (engendered perhaps by a reading of Gassendi or Locke) about our ability to know the essences of things. In a draft for the General Scholium he is emphatic in his assertion that we have no more idea of the "substances and essences" of things than "a blind man has of colors":

> We do not know the substances of things. We have no idea of them. We gather only their properties from the phenomena, and from the properties what the substances may be. That bodies do not penetrate each other we gather from the phenomena alone; that substances of different kinds do not penetrate each other does not at all appear from the phenomena. And we ought not rashly to assert that which cannot be inferred from the phenomena.[23]

Here he is questioning whether impenetrability could ever be known to belong to the essence of body. Most of this discussion was omitted from the published version of the General Scholium, but the direction of his thought is still clear: the "inner substance" of things cannot be discovered by any "reflex act of our mind".[24] 'Essential' carried for him, as for so many in the dualist climate of the time, the overtone of hiddenness, and he most emphatically did not want to claim this for his "universal" qualities which (he insisted) were fully open to sense.

A second objection to the term 'essential' derived from the debate regarding the status of gravity. If such qualities as impenetrability are called "essential", it is difficult to deny this title to gravity, since there seems to be no significant difference of standing between them (as both Newton and Cotes observe).[25] Yet he felt constrained to assert (as we shall see more fully in Chapter 3) that gravity cannot be "essential", nor (for the same reasons) can it be "innate".[26]

He *could* concede that it is "universal" (and allow Cotes to call it "primary") because for him that did not imply that it is grasped in its full physical reality, only that the phenomena distinctive of gravitational motion are associated with all bodies. Thus for him to call extension, mobility, and the like "universal" qualities meant that they are necessarily predicable of all bodies. But this was not a sufficient condition for their being described as "essential" in the stronger Lockean sense (nor, for that matter, in the different sense in which Aristotle spoke of essence). Universality was in fact the strongest claim that could properly be made for a predicate in Newton's system.[27]

§1.2 Intensity-Invariance as a Criterion of Universality

Though Newton does not in this context call upon the distinction between "primary" and "secondary" qualities he employs occasionally elsewhere, we can recognize in Rule III the conventional list of primary qualities found in writers of the day. He is distinguishing between qualities that are necessarily universally possessed and those that are not.[28] Yet his grounds for drawing this distinction are novel.[29] He goes to the terminology of the doctrine of "latitude of forms", inherited from fourteenth-century philosophy, to find the criterion for universal qualities, his argument being simply that such qualities "as are not liable to diminution can never be quite taken away" and are thus automatically recognizable as universal.

But to say that they are *invariant* is by no means to say that *all* bodies, even all bodies of our experience, possess them. Would it not be possible for some bodies to possess some of the invariant properties, such as mobility or impenetrability, and others to possess different ones? So he is forced to add a second criterion: that they also "should be found to belong to all bodies within the reach of our experiments", and then he makes use of the "analogy of Nature", to which he so often has recourse, to conclude that in this event they can be supposed to belong to all bodies whatsoever.

Newton proposes Rule III as his way of determining which are the universal properties of body or matter; it is of significance to us, then, because it represents his most persistent attempt to define materiality. But the puzzlement it has generated over the years warns us to take a closer look. There are two separate issues; one has to do with the criterion of intensity-invariance itself and the other with the apparent fallacy involved in extending a knowledge of the part to an unqualified assertion about the whole. This latter issue will be treated in the next section.

Why *did* Newton choose this criterion of intensity-invariance? As we have seen, he appears to have been led to it by his speculations about universal transmutability, and his realization that there had to be a limit to transmutability if bodiliness itself were not to be threatened by the incessant changes of matter. It was plausible to suppose that this limit would be set by the existence of material properties which were *not* transmutable. And to someone familiar with the doctrine of the "latitude of forms", it might well have also seemed plausible that such properties would be most readily identifiable by the criterion of intensity-invariance. After all, one does not need to know much about a property to decide whether or not it is intensity-invariant. There is no likelihood of an inductive error here. All one needs is to know what impenetrability *is* (and a relatively limited experience of bodies can tell one that) in order to see that it is not subject to intensity changes.

But let us examine more carefully the qualities he lists as satisfying the criterion. By 'extension' he presumably does not mean the particular extension a body happens to have (which is variable, and could hardly be called intensity-invariant) but rather its quality of being extended. Likewise, instead of speaking of motion, he speaks of mobility. In an earlier draft of the Rule, he had even implicitly underscored the contrast by saying that motion (*motus*) can be intended and remitted, whereas mobility (*mobilitas*) cannot.[30] The inertia he mentions cannot be a specific *vis inertiae* (which would not be universal); it is just the property of possessing inertia.

But now the language of intension and remission no longer seems applicable. The philosophers of the thirteenth and fourteenth centuries who constructed it with such elaborate care were quite explicit in restricting it to changes in intensive *quality*, and only to those continuous changes in quality which could be represented metaphorically by a line segment. Even in the domain of a quality (heat, for example), they distinguished between intensive changes in temperature and extensive changes in the total quantity of heat; it was only to the former that the terminology of intension and remission could be applied.[31] The purpose of the terminology was to use the resources of geometry to analyze changes which, although they were not strictly extensive in nature, still lent themselves to a convenient quantitative metaphor.

Extension, of course, is not a quality. But Newton speaks of extendedness, the property of being extended, which does not (he says) admit of degrees, since one must either have it or not have it. Is that a quality? In medieval discussions of the role played by quantity in the individuation of material beings, the distinction between the general property of extendedness (*dimensio interminata*) and the specific quantity of a particular extension (*dimensio terminata*) was very carefully drawn.[32] Neither property belonged to the category of quality—that was clear. The language of intension and remission was inapplicable, therefore, to them. It is not that extendedness is intensity-invariant; it is a category-mistake to apply this language to it in the first place.

But why could not Newton use the older terminology to a new purpose? It is not intuitively meaningful to say that extendedness does not admit of degrees? The answer, of course, is yes, but it does not, alas, help. For the same would be true of *any* second-order property, for example, the property of possessing color or weight. Color or weight may be variable, just as extension is. But the property of being colored does not admit of degrees: an entity is either colored or it isn't. In short, whether or not a property X admits of degrees, the *possession* of X does not. This is a purely conceptual point, deriving from the two-valued character of the propositional form "*S* is *P*." So that Newton's criterion not only strains the language of intension but is so broad as to be useless. All the weight has to be borne by the second criterion specified by the Rule: for a property to be a "universal" quality, it has "to be found to belong to all bodies within the reach of our experiments".

In the light of these difficulties, one may well ask whether Newton makes any attempt to *apply* the intensity criterion to rule out some properties and accept others. A glance at the discussion paragraphs that follow the Rule in the text shows that the criterion is not so much as mentioned. He discusses his candidates for universal status: extension, hardness, impenetrability, mobility, and inertia one after another; but in no single case does he allude to the intensity criterion. This is all the more striking in that some of these properties—hardness, for example—might seem to violate the intensity-invariance requirement right away. The criterion he applies to them is in each case the same (except for hardness; see §1.4). He asks whether our senses tell us that the property in question belongs to all the bodies within the range of our experience, nothing more.

Newton, therefore, could have omitted the troublesome intensity criterion from the published version of Rule III, without in the least affecting his manner of applying the Rule to concrete cases. That he retained it testifies to the important role it had played in the earlier draft, where he had thought he had hit upon a logically cogent way of singling out the universal qualities of bodies: if a quality is not subject to remission it must be universal.

> And the reason is that a quality which cannot be remitted cannot be taken away from the whole; while on the other hand, that which can be taken away, if it is taken away from some parts of the whole, it is remitted in the whole.[33]

Only in the context of the discussion of universal transmutability could this have attained the plausibility it briefly possessed. At most, it proves permanence, not universality, unless some intuitive notion of potential transmutability is employed. Doubtless, by the time the final text came to be written, Newton had become aware, in part at least, of the pitfalls involved. This would explain why the entire weight of Rule III was shifted to the second criterion, to which we now turn.

§1.3   Induction and the "Analogy of Nature"

Logicians have been wont to cite Newton's use of Rule III as a typical fallacy of composition ("the bodies that we handle we find impenetrable, and thence conclude impenetrability to be a universal property of all bodies whatsoever"). There can be no doubt—quantum theory would be the most spectacular counter instance—that the rule would now be regarded as an inadmissible maxim of scientific procedure. But was this so obvious in Newton's time? What sort of justification could he claim for it? He was not one to lay himself open to attack on the grounds of obvious fallacy. The beginnings of an answer may be found in the broader metaphysical and theological context of Newton's thought. Induction for him (as for most pre-Humeans other than the fourteenth-century nominalists) was in the first instance a method rather than a problem. Following his oft-quoted injunction in the General Scholium (1713) against the feigning/framing of hypotheses, he indicates the proper mode of procedure in experimental philosophy:

> Particular propositions are inferred from the phenomena, and afterwards rendered general by induction. Thus it was that the impenetrability, the mobility, and the impulsive force of bodies, and the laws of motion and of gravitation, were discovered.[34]

It sounds here as though he has a method of generalization sufficiently secure in its operation to warrant the certainty he attaches to the laws of motion. How is this "method" to be characterized? To answer this, it will be necessary to look more closely at a tangled cluster of questions that seventeenth-century theorists of science began to worry about in regard to the reach of the knowledge-claims open to science.[35] The first question was not a new one; medieval logicians had discussed it under the heading of *ampliatio*. How can one validly move from the particular to the universal? More specifically, if one's experience is of the Yness of only a limited set of Xs, how can one assert the Yness of *all* Xs? In the empiricist framework of most science of Newton's day, it was assumed that a knowledge-claim can have no other source than experience, and such experience is limited, evidently, to a finite range of the possible instances of the predicates involved. We can call this the "problem of induction," even though we are speaking of the century before Hume came along to underline just how problematic such inference is, if empiricism be pushed to the extreme he favored. Note that "induction" is restricted here to inferences from some (experienced) members of a presumptive class to other/all (unexperienced) members of the *same* class. It is the procedure emphasized by those who see science as laws or generalizations. It must be distinguished from what Peirce would call "retroduction", where one moves from instance to explanatory hypothesis and back again, where the hypothesis introduces new concepts or constructs and is not merely a generalization involving the same more or less descriptive categories as the instances from which the inference begins.

There is one particular form of induction that causes special distress to the empiricist: where one moves from particular instances which are observable to instances which are *in principle* unobservable. If the only legitimate ground of knowledge is sense experience, how can one justify a claim which cannot, *in principle*, be tested directly against experience? Yet the most striking feature of seventeenth-century science was its move into the "invisible realm," its introduction of a form of structural explanation that Aristotelian science had excluded. The corpuscular philosophy of Galileo, Boyle, Gassendi, and the rest postulated corpuscles which could be described in the same terms of extension, impenetrability, etc., that we would apply to objects of ordinary experience, even though these corpuscles, in principle, can never be experienced by us. There were those, like Locke, who worried about the validity of such a move, but most scientists incorporated the new "corpuscles" into their physics without the addition of any new concepts. No new language was needed for them; the "primary" qualities on which science would rest were those which the corpuscles shared with the sensible bodies of our daily experience.

This is a special form of *induction*, then, since it moves from particular instances to other instances *of the same kind*, that is, characterizable by the same predicates and covered by the same laws. McGuire seeks to distinguish it from induction and coins the term 'transduction' for it.[36] We will retain this useful term, but will treat transduction as a particular type of induction. To be sure, transduction poses a special problem for the empiricist. But the *logical* issue of moving from an assertion about particulars to a universal claim about a class as a whole is the same in each case. In both, it is in principle impossible to test all the instances of the universal; that the source of the impossibility lies at a deeper level for transduction than for induction generally does not seem to furnish an adequate ground for stipulating that the former is not a species of the latter.

Induction and transduction cover the cases where no new concepts are involved, where the homogeneity of the class is assumed and the categories are univocal. But what if our original concepts, describing our starting point in ordinary experience, are inadequate for the task of describing the term to which our argument is attempting to progress? We may employ analogies, perhaps, and assume that our starting concepts have to be modified in some way in order to make an assertion about the further particulars beyond those already experienced. Or, more radically, we may shift from generalization to hypothetical explanation, and introduce entirely new concepts whose function is to explain the empirical instances from which the inquiry began (retroduction). What these moves serve to underline is a shortcoming of *language*: concepts appropriate to an empirical starting point in observation may be only partially adequate, or perhaps entirely inadequate, when it comes to formulating the scientific "universal" that these observations warrant. Let us call this the "problem of transdiction", therefore, appropriating a term of Mandelbaum's to a new use.[37] Transdiction requires one to move from one mode of speech to another, and the problem of transdiction lies in justifying such a move.

Having defined three sorts of issue that arose for seventeenth-century scientists in their attempts to reason validly from a limited empirical starting point to a set of general claims about Nature, let us see how Newton responded to them and, in particular, which of them was involved in his use of Rule III to validate the list of the "universal qualities" of all material things. Newton, as we have seen, spoke often of the role played by "induction" in "making general" the particular claims of observation and experiment, especially in regard to the laws of motion on which the structure of the *Principia* rests. These latter, he says to Cotes, are *not* hypotheses; they have been "deduced from phenomena and made general by induction, which is the highest evidence that a proposition can have in this philosophy".[38]

Book III of the *Principia* ("The System of the World") applies these laws (whose mathematical consequences were explored in Books I and II) to the "phenomena" of the planets; the success of this application indicates the generality of the induction attained and points to the universal applicability of the laws to all matter.[39] Rule IV, at the beginning of Book III, gives some guidance in regard to the weight to be attached to the propositions he establishes here:

> In experimental philosophy we are to look upon propositions inferred by general induction from phenomena as accurately or very nearly true, notwithstanding any contrary hypotheses that may be imagined, till such time as other phenomena occur by which they may either be made more accurate or liable to exceptions. This rule we must follow, that the argument of induction may not be evaded by hypotheses.[40]

There is a slight hesitation in regard to universality here; he is prepared to admit that small modifications might have to be made in the light of later observational evidence. But the substantial applicability of the laws to the planetary system, and to bodies generally, is not in doubt.

But what of the more hazardous inference to the unobservable corpuscles of which matter is composed? In Query 31(23) he speculates:

> God is able to create particles of matter of several sizes and figures, and in several proportions to space, and perhaps of different densities and forces, and thereby to vary the laws of nature and make worlds of several sorts in several parts of the universe. At least, I see nothing of contradiction in all this.[41]

This cautious estimate was a consequence of his speculative investigation of the possible varieties of forces acting between corpuscles in the "invisible realm", forces that would explain chemical, electrical, magnetic, and other phenomena. But this caution would not have extended to the notion of force itself as defined in the three Axioms of the *Principia*. Nor did it deter him from making flat assertions about the qualities possessed by the primordial particles and by macroscopic bodies alike. And he quite clearly realized what was involved in this move. Unless it can be made with assurance, there could be no way of knowing that the mechanics of the *Principia* applies to the ultimate constituents of bodies. Indeed, there would be no way of knowing these constituents at all, nor *a fortiori* of using them to explain the macroscopic secondary qualities of things. In an early draft of Rule III, he says that it is "the foundation of all philosophy". Without it, "one could not derive the qualities of insensible bodies from the qualities of sensible ones".[42] Thus Rule III is basically concerned with transduction.

This can also be seen quite clearly in the applications he makes of it to argue, for example, from the impenetrability of the bodies of our experience to the impenetrability of "all bodies whatsoever". In the context, this is to apply in particular to the primordial particles, forever in all likelihood below the range of observation. The concept of impenetrability is to be univocally applied here; there is no question of analogy, that is, of a shift in the meaning of the concept when applied to the invisible realm. Such a shift would endanger the applicability to that realm of the basic mechanical framework of the *Principia*, and Newton could not envisage this as a serious possibility. The rule is *not*, therefore, one of transdiction (in the sense defined above). It relies on univocity and homogeneity rather than on analogy and continuity.

What about Newton's use of the phrase 'the analogy of Nature' in the commentary on the Rule? McGuire rests his case for transdiction on this phrase and on the transmutability context in which the Rule originated. He argues, against Mandelbaum (who makes the Rule a matter of transduction only), that the metaphysics in which the Rule is grounded is that of the great chain of being, the *scala naturae*, the steps and gradations from one order of being to the next, and not one of the uniformity of nature, as Mandelbaum would have it. Now there can be no doubt that Newton *does* speak at times of a continuous spectrum of beings, from the level of pure spirit to that of totally passive matter.[43] He also makes frequent use of analogies in his physics, for example, in extending the notion of vibration from the domain of acoustics to optics.

This, however, does not permit us to interpret the 'analogy of Nature' phrase in Rule III as a reference to the classical metaphysics of analogy or to the neo-Platonic metaphysics of the *scala naturae*. In fact, this is precisely what Newton, in the successive drafts of the Rule, was trying (it would seem) to exclude. It was to restrict the universal transmutability of all properties, to single out certain universal properties that were *not* subject to gradations, that Newton devised the Rule in the first place. The metaphysics of analogy that played such an important part in medieval philosophy was concerned with the problem of predication about God. None of the predicates of ordinary human language can be properly applied to Him; so how can *any* claim, even one of existence, be made about Him? Aquinas constructed an intricate theory of predication within which an *analogous* attribution of being to God can be properly made. There were, however, no "universal qualities" according to this view.[44] Likewise, the *scala naturae* requires a shading-off of such qualities as activity, as one moves down the spectrum. There is a *continuity* from one level to the next, and this continuity permits one to hazard analogical extensions of predicates upward or downward. But the very notion of a *scale*, of a hierarchy, would not support a straightforward univocal application of the *same* predicate to entities at different levels.[45]

What, then, *does* Newton mean by introducing the phrase 'analogy of Nature' into a context which is so emphatically univocal? He provides us with a helpful gloss: Nature, he says, is wont to be "simple and always consonant with itself". This "consonance" is clearly one of uniformity or homogeneity: the *same* is to be expected in the domain into which the analogy is extended. If all the bodies we experience are impenetrable, then *all* bodies may be expected to be impenetrable. In the *Opticks*, a similar passage occurs: the fact that macroscopic bodies act on one another by means of various sorts of attraction indicates the "tenor and course of nature" and suggests that additional kinds of attractive force may be found to operate at the level of the primordial particles, "for Nature is very consonant and conformable with itself".[46] There is no suggestion of a hierarchy or chain here; the course of Nature leads one to expect a variety of attractive forces at the level of the corpuscles, of the kind one already finds at the macroscopic level. Even more explicit is a passage in a draft for the *Opticks*: "And if Nature be most simple and fully consonant to herself, she observes the same method in regulating the motions of smaller bodies which she doth in regulating those of the greater".[47] Note: the *same* method. And he goes on, significantly, to add that this "principle of Nature" is "remote from the conceptions of philosophers", perhaps because philosophers would have been likely to prefer analogical to univocal language in contexts such as these.

On what, finally, does his use of transduction rest? Two related supports may be cited. One was his notion of a cosmic order which is simple and consonant with itself; this in turn was rooted in the traditional Christian theology of Creation. The second is, as we have seen, not so much a support, strictly speaking, as a regulative principle: without something like Rule III, Newton did not see how science itself would be possible. There *had* to be a way of extending mechanics into the invisible realm; it was a condition of the possibility of science that Nature be of such a kind as to permit this:

> Nature would not be sufficiently simply and consistent with itself [unless] the qualities of bodies on which it is possible to make

experiments, as many of them as are found to be immutable, apply equally to all bodies.[48]

§1.4   The Definition of Matter

Rule III provides a handy definition of matter in terms of the familiar primary qualities of earlier seventeenth-century thought; in the final version of the Rule, Newton somewhat hesitantly adds *attraction* to them. Of its appearance here, we shall have more to say later (§3.4). Matter is no longer the indeterminate co-principle of substance it had been in Greek and medieval philosophy. Now it defines, in a quite concrete way, the range of applicability of the concepts and methods of the new "experimental philosophy". To be material is to be extended, solid, mobile, to possess inertia, to attract, and be attracted by, all other bodies. It is, therefore, to be subject to the laws of mechanics. In earlier drafts of Rule III, indeed, he hesitated, when claiming universality, between "qualities" and "laws":

> The laws and properties of all bodies on which experiments can be made, are the laws and properties of bodies universally.[49]

In the prefatory remarks of Book III, Newton tells the reader what he is doing:

> In the preceding books I have laid down the principles of philosophy, principles not philosophical but mathematical: such, namely, as we may build our reasonings upon in philosophical inquiries. These principles are the laws and conditions of certain motions, and powers or forces, which chiefly have respect to philosophy. . . . It remains that from the same principles I now demonstrate the frame of the System of the World.[50]

The "Rules of Reasoning in Philosophy", which come next, are evidently intended to aid in this application of the mathematical principles (enunciated mainly, he says, in the

Definitions, Laws of Motion, and the first three sections of Book I) to the "philosophical" task of discovering the planetary motions. One must above all be able to assume the universality of the properties which serve as necessary and sufficient conditions for the applicability of the Definitions and the Laws. This is evidently one reason why Rule III was needed.

A second, as we have seen, was to mark off Newton's concept of matter from that of the Cartesians. Since Book III is focused on gravitational motion. it was important to discount the main rival to his own view, which was the aether-vortex theory of Descartes. One way of doing this was to show that the equating of matter and extension, on which the aether theory rested, is untenable: "all spaces are not equally full".[51] Though all matter is extended, not all extension is material. Besides, matter is also hard, possesses inertia, and gravitates; none of these properties is compatible with the Cartesian notion of matter-as-extension. Nor are they properties of Descartes' aether. That aether, therefore, cannot be counted as material, and cannot in consequence be a proper subject for "experimental philosophy" either.

Does Rule III *really* serve to generate the list of universal qualities Newton specifies? Had he not decided on this list beforehand? Could Rule III, as it stands, certify hardness or impenetrability as universal qualities? It is significant that, when discussing hardness, instead of saying that *all* the bodies of our experience are hard (the form he follows for the other qualities), he simply remarks that an "abundance of bodies" (*corpora plura*) are hard, and that such hardness can be explained only by postulating the hardness of the undivided particles of which they are composed. From this, in turn, it can be inferred that all undivided particles are hard. Note that this is an argument from reason ("we justly infer"), a theoretical argument. And it construes the notion of universal quality as applicable only to the primordial particles, not necessarily to all the bodies constituted by them (some of them are clearly *not* hard).

When he goes on to impenetrability, however, he asserts: "that all bodies are impenetrable we gather not from reason but from sensation. The bodies which we handle we find impenetrable...". But surely this will not do. Newton himself frequently insisted on the vacuities that permeate the bodies of our experience. Despite their appearance of solidity, he came to believe that these bodies are "rarer by far than is usually believed". For example;

> Gold is not solid; it abounds in pores. For it is dissolved by penetrating acid waters, and quicksilver entering through its pores easily pierces its interior parts and whitens them to the very center.[52]

So that even gold, solid as it is, is not really impenetrable. It is difficult to reconcile this with the assertion in Rule III, which he repeats elsewhere:

> Another principle is that matter is impenetrable by matter. This is usually looked upon as a maxim known to us by the light of Nature although we know nothing of bodies but by sense. We find by daily experience that bodies resist one another as often as they come together, and cannot by any force be made to penetrate one another's dimensions.[53]

He goes on to infer, as in Rule III, that since impenetrability is a property of all the bodies we observe, it must be a universal property of matter; "and such observations occurring every day to every man, this property of bodies is acknowledged by all men without dispute". He is trying hard, against the odds, to make impenetrability an experienced property of experienced bodies, rather than a property of body derived by the "light of nature", the *lumen naturale* of rational reflection so much relied on by Descartes.[54]

Rule III evidently hesitates between two very different epistemologies. First is the empiricist one, which would regard the "universal qualities" as sense qualities that all the bodies we observe share in common (like the "common sensibles" of the Scholastic tradition). They would thus be defined as they are given in perception. The persuasive force of the Rule would lie in the inductive move: because all the bodies we observe share quality Q, it is plausible to suppose that *all* bodies (including unobservable ones) possess Q. This inference depends squarely upon the homogeneity principle discussed above. Its strength lies in its direct inductive appeal; its weakness lies in possible challenges to the homogeneity principle, and also in the fact that at least two of the qualities his application of the Rule singles out for universal status are, fairly obviously, not in fact common to all bodies of our experience.

As we read these passages, a quite different epistemology also shows itself. It is indicated by the use of the term 'experiment' (rather than 'experience') in the Rule, and in the occurrence of theoretical argument for the necessity of matter's exhibiting a particular property (rather than inductive inference from perceived instances). Attraction is one of the universal qualities; indeed, it is attraction that he is probably most anxious to qualify as universal. Yet it is clearly not given us in experience, as, for instance, the mobility of body is. It "appears by experiments and astronomical observations", he says, that all bodies gravitate toward the earth, and that "our sea gravitates toward the moon". But clearly there is a highly interpretive scheme involved here. It is not *attraction* that is perceived in the bodies of our experience; it is the regularity with which they fall. But to see this as "attraction", that is, as an instance of the same force as that responsible for planetary motion, requires the entire theoretical structure of the *Principia*. Attraction is not given directly by experiment but only by the theory that experiments and sophisticated astronomical observations suggest. This is a far cry from the simple inductive appeal of the "straight" Rule.

Newton almost certainly did not realize how far from inductivism his attempt to include attraction among the universal qualities leads him. But the argument he gives for the inclusion of hardness, as we have seen, relies not upon the universality of hardness among all the bodies we observe but rather upon the necessity of *postulating* hardness as a property of the (unobserved) primordial particles of the solid bodies we do perceive (the argument being that only the hardness of the parts would explain the hardness of the whole), and then of inferring that all *other* primordial particles share the same quality. The structure of this inference is not inductive: the primordial particles of the hard body are said to be hard not because hardness is a universal quality but because only thus can the hardness of the observed body be *explained*. And we do not rely on the claim that all observed bodies *are* hard.

The methodology in both these instances is a retroductive one. And the universal quality in the second case is postulated only of the primordial particles. Only they are held to exemplify body or matter in its fullest sense in this instance. Newton evidently wishes to distinguish his method from the conceptualist one, according to which matter would be said to be impenetrable because the concept of matter, properly understood, necessarily involves the concept of impenetrability (a maxim "usually looked upon as known to us" by the natural light of reason).[55] Yet there are echoes of a conceptualist method in the retention of hardness and impenetrability, which do not appear to be needed for his main purpose (to assure the universal applicability of the "mathematical principles" of Book I),[56] and which cannot provide a basis for the sort of inductive argument Rule III purports to describe. The most plausible way to represent matter as hard/impenetrable/solid is to take 'matter' to refer to the primordial particles and to suppose that they must be mutually impenetrable because it would be contradictory to suppose... etc.

What kind of definition of matter do Newton's "universal

qualities" finally afford us? It is not strictly empiricist; the qualities are not entirely given us in observation (though the degree of "theory-ladenness" of qualities like inertia and extension would not be revealed until nearly two centuries later). It gives us more than the strict minimum required to qualify matter as the proper subject of the experimental philosophy of the *Principia*. The definition is also implicitly reductionist: the universal qualities provide a sufficient characterization of matter to explain the entire range of behavior of physical bodies. Secondary qualities, such as color, are real, but are not more than dispositions, whose ontological basis in physical objects can be understood in terms of the motions and relations of primordial particles.[57] One need involve no qualities other than the minimal list of Rule III for the broad purposes of natural science, here conceived of as a sort of generalized mechanics.

Nonetheless, Newton is more careful than his seventeenth-century predecessors had been about the reductive sufficiency of his definition of matter in terms of the universal qualities. He never says that all other properties can be derived from them; he says that with them *and* a variety of forces and active principles, only some of which are as yet known, all natural phenomena can be explained. Newton's analysis of inflection in optics had proved to his satisfaction that force has to be admitted as an irreducible category in physical explanation; the success of the *Principia* amply bore out this conviction. But the addition of force to the repertoire of mechanics basically altered the problem of reductionism.[58] No longer was it plausible to suppose that, in principle, all changes could be explained by specifying a restricted set of mechanical properties. Laws of force have to be specified too, and there is no way of knowing how diverse they will be or of knowing that all of them have been discovered.

Newton's own hopes for his concept of matter were, of

course, quite naturally reductionist, but he was very tentative in expressing them:

> To derive two or three general principles of motion from phenomena, and afterwards to tell us how the properties and actions of all corporal things follow from these manifest principles, would be a very great step in philosophy, though the causes of those principles were not yet discovered.[59]

CHAPTER TWO

Is Matter Active?

CAN MATTER BE said to be of its nature active? It might have been supposed that the principle of universal gravitation, from which the success of the *Principia* so obviously derived, would have led Newton to give a resounding affirmative to this question, or, to put it the other way round, that a belief in the inherently active role of matter would have helped him to the notion of gravitation. But, in fact, the opposite seems to have been the case. The philosophical traditions that most influenced the young Newton disposed him instead to the view that "matter is a passive principle and cannot move itself", a thesis that recurs frequently in his writings.[60] It often seems that "matter" was for him, almost by definition, the passive principle of mechanical systems, "that which is moved". If something possesses within itself a source of activity, it cannot *just* be "matter", he seems to say; there must be another principle, an "active principle" distinct (though not necessarily physically separate) from the matter involved.

Newton felt himself forced to look outside matter for the source of motion, and it was this, perhaps, more than any other factor, which gave his ontology such complexity. He was not satisfied with any of the answers he found, and in consequence he kept modifying his ontology throughout his life.[61] It may be instructive to contrast his general position in this regard with the Cartesian and Leibnizian natural philosophies, the two views that he was continually forced to distinguish

29

from his own. Indeed, the effort to separate his concept of matter from that of Descartes at one extreme and that of Leibniz at the other clearly dominated his mind in the 1690s and again in the 1710s, as a reading of the draft material from these periods shows.

Descartes proposed a two-level ontology of matter and spirit, with motion an ambiguous third element.[62] Matter was essentially passive; motion was communicated to it at Creation and the quantity of motion was thereafter conserved. A major difficulty with this position was that it provided for no creative natural source of motion in the physical world; it could not, therefore, account for accelerated motion (fall, planetary motion, etc.) without invoking imperceptible and ultimately implausible vortices. Yet this view of matter seemed to follow with inexorable logic from the Cartesian method itself, which demanded an ontology of geometrically intelligible entities, that is, volumes, that obviously could not be regarded as active in their own right.[63]

§2.1 Leibniz

In his "new science of dynamics" (first formulated in 1686), Leibniz emphatically rejected this Cartesian ontology of an inert "matter" because it appeared to him to leave natural agency entirely unaccounted for. On the contrary, he argued, matter is intrinsically active: "the whole of Nature is full of life".[64] There are two levels of approach to the world, one metaphysical and the other phenomenal. At the former level, each monad is self-sufficient; its concept contains all its predicates, past, present, and future, and hence it may not be acted upon by anything outside itself. Its "primitive force" is conserved; it is both active, grounding the activities of substance at the phenomenal level, and passive (as *materia prima*), grounding the phenomenal manifestations of extension, impenetrability, and resistance. From the phenomenal standpoint, the standpoint of the science of dynamics, matter is suffused with (derivative) forces, again both active and passive.[65] The motion of (phenomenal) bodies is due to active forces, internal to the bodies themselves; the resistance they manifest to change of motion is a passive force. Bodies are *not* indifferent to motion; they will respond in characteristic ways to attempts to change their state of rest or motion. The heavier they are, the more resistance they show. Their "natural inertia" is proportional to their quantity of matter or to their weight. (He did not distinguish between these, nor did he suggest that the inertia of matter could be responsible for the continuance of uniform motion. In his view, the passive force of inertia acted *against* change of motion. It would have made no sense to him to attribute any sort of motion directly to it as cause.)

The active forces are of two sorts, *vis viva* (force of motion) and *conatus* (also sometimes called *vis mortua* or *solicitatio*). *Vis viva* is the proper measure of the force possessed by a body on account of its motion, the measure of what it can accomplish; it is expressed as mv^2 (not, as mv, the measure proposed by Descartes).[66] *Conatus* (called "dead force" because it does not itself arise from motion) is responsible for *changes* of motion; were it not for this "dead force", the world would rapidly run down, because *vis viva* is lost to macroscopic bodies at imperfectly elastic impact. (The total sum of *vis viva* in the universe does not diminish, since it is taken up by the small parts of which bodies are composed.)

All action is contact action, though, strictly speaking, a body which strikes another is an *occasion* for the active forces internal to the second to manifest themselves, rather than a *cause* of its consequent motion. Leibniz agreed with Descartes in excluding the possibility of void space, but his reasons were very different from those of the earlier philosopher. He argued that only a *plenum* (at the phenomenal level) is consonant with God's infinite power: "for the perfection of matter is to that of a vacuum as something is to nothing".[67] Indeed, *every* created substance is "accompanied by matter". Since space obviously plays a role in mechanics and optics, it cannot be regarded as non-being. Yet it is not a reality in its own right either. It can properly be described as "material" since it must contain matter (he mentions light, magnetic force, and aether) throughout.

§2.2   Newton

It is interesting to note the analogies between this system and that of Newton, and yet the profound philosophical differences that underlay the two theories of motion. Newton's position in regard to the activity of matter is intermediate between those of Descartes and Leibniz. Like Leibniz, he saw the insufficiency of the Cartesian ontology in accounting for any motions save inertial ones. By including force as an additional ontological element, he succeeded in transforming what Descartes had left as explanation sketches into testable predictive accounts. But right from the beginning, his notion of force exhibited the same troublesome ambivalence between activity and passivity as that of Leibniz. The role it played in the explanation of motion, however, was quite different in the two. Where Leibniz made *vis viva* primary, Newton stressed *vis impressa* (which corresponded with the *vis mortua* of Leibniz, that is, that which initiates new motion). Indeed, Newton's principle of inertia excluded *vis viva* from the realm of active force entirely, because uniform motion, in his view, neither required nor of itself constituted such a force. (It took a long time for the relations between these different conceptions of force—cause of change of motion and ability to perform work—to be clarified.)

On the side of "passive force", however, there was a remarkable similarity between the two systems. In each, there is a "force of resistance", proportional to quantity of matter, somehow resident in every body. This fits neatly into Leibniz' general account of matter as dynamic; one of the characteristic dynamisms of matter is its effort to remain in the state in which it is. But it introduced an incoherence into Newton's theory from the start, because of the attribution to a supposedly inert matter of something called "force", and the proportionality of this *vis inertiae* and mass, itself inexplicably determinative of *vis impressa*, and thus, apparently, a measure also of the capacity of a body to act on another body.

§2.3 *Vis Inertiae*

The very first occurrence of the word 'force' in the *Principia* (Definition III) suggests how intractable this problem will be for Newton:

> The *vis insita* or innate force of matter is a power of resisting, by which every body, as much as in it lies, continues in its present state, whether it be of rest, or of moving uniformly forward in a straight line. This force is always proportional to the body whose force it is, and differs nothing from the inactivity of mass, but in our manner of conceiving it. A body, from the inert nature of matter, is only with difficulty changed in its state of rest or motion. Upon which account, this *vis insita* may by a most significant name be called the force of inertia [*vis inertiae*]. But a body only exerts this force when another force impressed on it endeavors to change its condition. And the exercise of this force may be considered as both resistance and impulse: *resistance*, insofar as the body opposes the impressed force in order to maintain its present state; *impulse*, insofar as the body endeavors to change the state of another body by not easily giving way to the impressed force of the other. . . .

This passage brings out the tensions inherent in Newton's interrelated concepts of *matter* and *force*. Matter is said to be inert, not in the sense of total passivity but implying, rather, that "only with difficulty" is its state of rest or motion changed. In this respect, inertia connotes resistance to change of motion, and appears to be associated by Newton with the matter-aspect of bodies specifically.[68] But why is this called "force" (*vis*), since it "differs nothing from inactivity"? Of course, it *did* differ from inactivity (as the decision in the final version to use the odd phrase, '*vis inertiae*', showed). And it also differed from force, at least as force was defined in Law II. Bodies are said to exert an "impulse", furthermore; when their own state of motion is affected; they "endeavor to change the motion" of other bodies. Newton notes the obvious distinction between these two aspects of a body's reaction to impressed force; one he calls "resistance", the other "impulse".[69] But how is all of this to be put together consistently?

Before tackling this notoriously thorny issue, let us recall two other contexts in the *Principia* where the "innate" forces play an explicit role.[70] The first is Corollary I to the Laws of Motion, which treats of the composition of forces. At first sight, the "forces" mentioned here might seem to be forces in the modern sense, and the theorem the familiar one regarding the parallelogram of forces. But this is not the case. Each "force" is said to cause *uniform* motion; what are compounded, therefore, are two *impulses* at an angle to one another, each of them giving rise to a uniform motion. The starting point of this analysis is still (as it had been in the earlier *De Gravitatione* and *De Motu*) the notion of *impact*; change of motion is most obviously brought about when one body strikes another. The notion of a continuous force is far less obvious. Because its operation cannot so easily by tested for or treated mathematically, Newton has to work gradually toward it.

The second passage to note is Proposition I (the area–time relationship in a central-force orbit). Here Newton compounds the *vis insita* of the body with the *vis impressa* of the central force, taking both of them equally as sources of impulse (*impetus*) which can then be compounded by Corollary I. The modern reader is likely to be scandalized at the apparently carefree equating of momentum and force as causes of motion. The dimensionality, he notes, is wrong, and he immediately wonders about the validity of a theorem that seems to conflate first-order and second-order infinitesimals so flagrantly.[71] Even more striking, from the point of view of our inquiry into the concept of matter, is the attribution to matter of an "innate force" which is responsible for uniform motion, and which has an active role not only in impact but also in motion under central forces. This is a far cry indeed from the conventional textbook characterization of Newton's Laws as embodying the discovery that only *change* of motion needs a cause. Such a claim would be compatible with a reading of the three Laws in isolation from all the surrounding text. But it is untenable once this larger context is taken into account.

The concepts of *vis insita* and *vis inertiae* signaled the abandonment of the principle of the strict passivity of matter which had so heavily influenced earlier mechanics. But their obvious ambiguity was to give Newtonians many a headache in the century that followed.[72] And the headaches have not ceased: historians of science are deeply divided as to how the roles attributed to body/matter in the *Principia* ought to be understood and, in particular, whether they can be brought together in a single consistent mechanics. In this brief treatment, we shall not try to analyze the historical sources of Definition III but shall simply concentrate on the implications of the various notions of force in these first pages of the *Principia* for an understanding of Newton's complex and tentative concept of matter. But let us first lay aside the convolutions of those pages in order to distinguish, in a stipulative way, between three different sorts of "force", each corresponding to a different "activity" of matter. This may make it easier to trace our path through the laboriously constructed text, where Newton is still groping for the best terminology to convey what he wants to say.

(1) *Vis conservans* (VC) is a conserving "force", the cause (in some sense) of a body's continuing uniform motion. It is measured by the impulse required to bring the body from rest to its present velocity, or equivalently by the body's total momentum. It acts in the direction of motion of the body; it renews the body's motion at successive instants by a sequence of impulses.[73] It is, nevertheless, not a *vis impressa* in the sense of Law II.[74] It has overtones of the impetus of earlier physics: its measure is similar, and so is its function. But unlike impetus, it is not thought of as being superadded from the outside; it is an "innate force", a characteristic of any moving body in its own right. In an absolute space, it will presumably have an absolute value, namely, the momentum of the body with respect to absolute space. Nor is it continuously acting; rather, it is represented by a series of impulses, instantaneously creating and destroying the speed of the body.

(2) *Vis resistens* (VR) is a resisting "force" equal and opposite to the *vis impressa* (VI) that is responsible for its appearance. It is exerted only when a force acts to change the motion of a body; it is not operative when the motion is uniform. Its measure is that of the force imposed and not of the momentum of the body on which the force acts. It involves Law III. When A acts on B, B reacts. The reaction, considered as pertaining to B, is VR, whereas in regard to A it is VI since it changes the state of motion of A. The crucial point here is that resistance is understood to be ontologically responsible in some way for reaction.

(3) *Vis impressa* (VI) is impressed or external motive force in the standard sense of later mechanics. VI and VR appear to be two sides of the same coin, yet one has to be careful in expressing this. If VI and VR have the same sort of reality considered as force, paradox is imminent. For then there will always be

equal and opposite forces acting on a body, which would mean that no change of motion would ensue. [75] If action and reaction are equal, as Law III states, how can anything ever happen? The quick (modern) answer is that action and reaction occur in different bodies: if A acts on B, B reacts on A, so that there is no question of equal and opposite forces acting on the *same* body. But if VR is postulated, the analysis is not so simple. When A acts on B, the action (with respect to B) is a VI; the resistance to it in B is a VR. This resistance (in B) in turn is responsible in some way for the reaction (VI) on A, but is of course not identical with it. The operation of Law III is construed as involving not only two *vires impressae* but also the two correlative "forces of resistance" (VR), responsible for and at the same time elicited by the impressed forces.

Now let us return to the text of Definition III (see above) to see how these three strands are interwoven there. At first sight, it is tempting to equate *vis insita* with VC since its specified role is to maintain the body in its present state. And the emphasis in the *'vis inertiae'* label is surely on resistance to change of motion; so is it not equivalent to VR? What spoils this attractive reconstruction is that Newton says that *vis inertiae* and *vis insita* differ only in name; they are, apparently, the same thing taken from two different points of view. But this certainly cannot be true of VC and VR. To maintain a body in motion with a "force" proportional to the mass of the body acting in the direction of motion of the body is *not* the same as to resist change of motion with a "force" which is equal and opposite in direction to the impressed force.

Two different contexts appear to be envisaged in Definition III. In the description of *vis insita*, a body is assumed to be in uniform motion, whereas *vis inertiae* is said to be elicited only in non-uniform motion. *Vis insita* is somehow responsible for the continuance of uniform motion. But how? Its role is described in two rather different, and not easily reconciled, ways.

First, it resembles a VC in that it conserves the motion.[76] In Proposition I, it is represented (as we have seen) as though it were a "force" bringing about uniform motion by means of a series of impulses capable of being compounded with the impulses generated by an external force. This innate "force", acting in the direction of motion, is a force, however, only in the sense of being represented by impulse. Is this just a mode of representation for the convenience of the mathematical analysis in Proposition I? It might seem so, since it would be hard to attribute physical reality to the instantaneous creation and destruction of momentum implicit in its operation. Yet Newton gives no indication in the text that it should be regarded as a fiction. And the term 'innate force', on its face, appears to denote some sort of causal agency.

It should be recalled once again that the dominant context of discussion in the mechanics of the seventeenth century was *impact*.[77] Thus "force" had been most often regarded as something that a body possesses because of its motion. If impact is the sole means of altering motion, as the Cartesians and Leibniz insisted, then force is principally "force of motion". Perhaps one may view Newton's *vis insita* as a gesture to this tradition. But it would be risky to construe it as similar in function to Leibniz' *vis viva* or *action motrice*, because the ontologies of the two thinkers differed so fundamentally.[78]

Besides the function of conserving motion, *vis insita* is also given a second role, described as a "power of resisting" (*potentia resistendi*). What is meant by "resisting" here? The weakest sense would be "avoidance of change of motion". This sort of "resistance" would be a direct consequence of *vis insita* as a conserving factor (VC). If motion is conserved, change of motion is avoided. But something stronger than this seems to be intended. He explains *vis insita* by saying that "because of its inertia a body can only with difficulty be put out of its state of motion or rest". So the connotation is one of resisting change. Is the "power of resisting" a disposition, then, which is actualized when an impressed force acts to change the motion of the body? Newton appears to have this in mind when, in the second half of the Definition, he says that *vis insita*, now under the name *vis inertiae*, is "exerted" only when an impressed force acts to alter the motion of the body. Is this "resistance" connected with the "reaction" mentioned in Law III?

Newton draws our attention to two ways in which the "exercise" of *vis inertiae* can be considered. First, it is impulse (*impetus*) insofar as the body "endeavors to change the state of motion of another body by not readily giving way to the impressed force of that other". Second, it is *resistance* "insofar as the body opposes the impressed force in order to maintain its present state". Three elements appear to be distinguished by him in terms of their causal roles. One is the endeavor (*conatus*) to change the state of the other body; the *conatus* in B is attributed some sort of responsibility for the external force of reaction operating on A. The second is the not-easily-giving-way, or resistance, from which the *conatus* apparently derives. And the third is maintaining the present state, which is stated to be the consequence of the resistance. This would make resistance the ontologically basic aspect of *vis inertiae* from which both the *conatus* and the conservation features would follow.

We have seen that the last of these (VC) is given an autonomous role in uniform motion. How about *conatus*? Is it to be construed as an ontologically distinct property of the body acted upon? We are back in the sort of issue that surfaced so often in Scholastic philosophy and that was exemplified so clearly in the earlier career of the concept of matter.[79] Once one embarks on this sort of ontology, it is difficult to stop: the *conatus* or striving is not identical, as a principle, with the resistance.[80] Thus we now may have, it might seem, no fewer than four principles at work in a body being acted upon by another body: the external force acting upon it (VI), the resistance the body puts up (VR), the *conatus* to alter the motion of the other body, and the innate force responsible for conserving the present motion of the body (VC) by impulse. Though they are intricately interrelated, they are not identical as agencies.

Resistance is mentioned twice in Definition III, once where *vis insita* is said to be a *power* of resisting, a disposition possessed by a body in uniform motion, and later where *vis inertiae* is said to be a force of resistance elicited only when a body's motion is altered. Once again, then, can *vis insita* and *vis inertiae* be related as disposition and actualization?[81] There can be no doubt about the "actualist" language in which *vis inertiae* is described as a resistance acting against the imposed force (although, as we have seen, we must beware of giving it the status of motive force in its own right). The difficulty lies rather with taking *vis insita* to be straightforwardly dispositional; the conserving function attributed to it is constantly operative (according to Proposition I) so that it cannot be *purely* dispositional. *Vis insita* has two aspects, as we have seen; if the aspect of constant conservation (VC) be left aside, one might regard the other aspect, the power of resisting, as the disposition latent in bodies in uniform motion which is actualized in terms of *vis inertiae*, considered as resistance, when a body is acted upon.

Their relationship to the "maintaining of the present state of the body" is quite different, however, for the *vis insita*, taken as the source of impulse for the continuance of uniform motion (VC), is represented as the cause of uniform motion, whereas the *vis inertiae*, considered as resistance (VR), does not in fact maintain the state of the body. It is represented as *tending* to do so by opposing the external force, but this "tending" does not contribute to the analysis of the motion in terms of impulses. Thus it is not an actualization in the sense expected. The conservation and resistance aspects of Newton's causal analysis are differently related in the two types of motion, uniform and accelerated. The neutralization of an impressed force and the active maintenance of uniform motion are ultimately quite different ways of "maintaining the state" of a body.

To sum up, then, there are three main barriers to a consistent interpretation of the web of agencies Newton calls on in Definition III to explain motion. The first is his construal of *vis insita* (in Theorem I especially) as a conserving agency (VC), continuously responsible for the uniform motion of a body not acted upon by external forces. This allows him to represent the action of *vis insita* as a sequence of impulses in the direction of motion of the body. The implication is that uniform motion still needs an "explanation" of some sort, though admittedly not an agent force in the full sense. The tension between this conception, so reminiscent of the older theories of impetus, and the newer inertial ideas already implicit in the Cartesian laws of motion is evident.

Second is the responsibility attributed to *vis inertiae* for the reaction that alters the motion of the agent body. This is a legacy, in part, of the *conatus* tradition; the notion of an "endeavor to act", as a positive reality elicited by way of opposition to an impressed force, is prominent in Newton's earlier writings. But this appears to make *vis inertiae* a source of activity in its own right, contrary to the original metaphor of "inertia".

Third is the equivocal character of Newton's notion of "resistance", which in the context of *vis insita* appears to have the function of somehow conserving uniform motion (VC), whereas in the operation of *vis inertiae* it appears as one member of a fictive "equilibrium" of forces. When a body is said to "oppose the impressed force in order to maintain its present state" by the exercise of a "force of resistance", one is inevitably led to think of this latter "force" as opposing the impressed force in its action upon the body (i.e., as VR), and not just as a straightforward force of reaction (VI) on the *other* body. If the "resistance" were to be nothing more than the latter under another name, the notion of an "opposition", of an endeavor to restore the present state, would be lost, since the forces would be acting upon different bodies.[82] Newton takes Law III to commit him to an action–reaction pair in each body, and the ensuing dynamic complexity is a major source of puzzlement to the modern reader. The problem is at bottom one of ontology; once again we have an illustration of how the ontological issues that Newton tries so hard to suppress in the *Principia* nevertheless affect the structure of mechanical agency proposed there.

This complicated story may finally be summed up with the aid of our VC/VR/VI distinctions. Despite the fact that *vis insita* and *vis inertiae* are said by Newton to differ only in name, the dynamic functions he attributes to them are not quite the same. Further, each of them turns out to be uncomfortably dual in nature. *Vis insita* is proposed both as conserving motion (VC) and as a disposition to resist change of motion (i.e., a disposition to VR, conceived as actualization). *Vis inertiae* is proposed both as a resistance or reaction to impressed force (VR) and as an impulse that is responsible for motion in the other body.

Newton has thus modified the traditional doctrine of the passivity of matter in three different ways, corresponding to the VC, VR, and *conatus* aspects discussed above. All three have their "source in some corporeal passivity".[83] The first is an agency, innate to matter (VC), which keeps a body in motion with the motion it already has. Second, there is the resistance with which a body responds to any attempt to disturb its motion, this resistance being construed as situated in the resisting body (VR). Third, there is the *conatus* by which a body endeavors to alter the motion of another body acting upon it. All three are taken to be rooted somehow in the matter-constituent of things, in matter's tendency to conserve whatever motion it has and to impose motion by way of reaction in consequence of this resistance.

All three are proportional to quantity of matter, but none is measured by it alone. *Vis insita* is "always proportional to the body whose force it is", that is, to the mass; as VC, it is also proportional to the velocity of the body. VR, on the other hand, is measured—insofar as the notion of measure is appropriate to it—by the quantity of the impressed force, which of course will be proportional to the mass of the body on which the force is acting, the body in which the VR occurs. It will also be proportional to the acceleration produced, though this is not immediately evident from the form of Law II, as Newton wrote it, which does not treat acceleration directly.

It should be recalled that Newton was trying to reshape the entire conception of matter. It is not surprising, then, to find tensions and inconsistencies in his account. Had he merely wished to present a mathematical scheme, there would have been no problem. But what he aimed at was a physics, and ultimately a natural philosophy. The success of the mathematical scheme ought not lead us to overlook the ways of thinking about the world that made the formulation of such a scheme plausible in the first place.

§2.4 Active Principles

These ways of thinking were in part the outcome of the neo-Platonic doctrines he had learned as an undergraduate from Henry More. The Cambridge Platonists found it necessary to broaden the categories of the mechanical philosophy in order to allow for the regular operation of God in the world and for the endless variety of activities in living things. To the passive principle of matter they added all sorts of active principles; without them, they argued, animation and purpose could not appear. This argument was to be echoed by Newton over the years, as we shall see. But it was to his interest in alchemy that Newton owed his most intimate acquaintance with the turbulence of natural process, and from this acquaintance came the strength of his conviction about the presence in Nature of perpetually working active principles.

The vast store of alchemical notes that he left behind comes mainly from the two decades 1675–1695. These notes have only recently come under scholarly scrutiny, having been regarded for the most part by earlier Newtonian commentators as a foible best overlooked. Newton, we now know, read hundreds of alchemical texts, laboriously collated them, and attempted to penetrate their involved symbolisms. For example, the *Index Chemicus* which he compiled, has 879 headings and nearly 5,000 references to more than 100 authors, representing the entire range of the alchemical tradition.[84] The immensity of the effort can hardly be exaggerated. He was evidently seeking clues to the nature of chemical and vital processes and because the mechanical philosophy could not help him, alchemy seemed a likely place to look. He was already disposed to the hermetic idea of an ancient wisdom hidden from all but the initiate; his reading of Christian theology had led him to seek in the prophecies and the *Apocalypse* the hidden truths of history. Their decipherment must have seemed not altogether dissimilar from the unlocking of alchemical symbolisms. Newton obviously enjoyed the challenge of interpreting the arcane.

The alchemical literature he pored over was full of references to the active principles responsible for the transformations of matter. The alchemists' belief was that the matter of all things is one and the same, and that variety and activity alike come from the animating principles they disguised under code names: 'Diana', 'serpent', 'green lion', and so forth. They regarded the process of chemical change as a type of living process, to be understood in terms of generation and corruption, nourishing and purging. A "science" of such change could not be formulated in terms of mechanics, it seemed clear. Even if one were to discount the hermetic and allegorical aspects of alchemy, the store of accumulated knowledge of chemical change might suffice to convince any but the most reductionist Cartesian or atomist that some further principles of explanation would be needed than bare matter and motion. Newton, at least, was quite convinced.

His chemical experimentation, which occupied most of his time in the periods just prior to and just after the writing of the *Principia*, gave him extensive knowledge of chemical interactions. One can see him relying more and more on his own findings and less on the alchemical authorities as time goes on. And his findings are expressed, as a rule, in a relatively exact and prosaic language, far removed from the allegorical codes he had spent so much time trying to pierce. But the gradual transition from alchemy to chemistry in no way diminished Newton's conviction that the world is permeated by active principles of all kinds.

The duality of active and passive principles is muted in the *Principia*, as we have seen, but it is by no means absent. From where does new motion come in this new system of the world? In his account of *vis inertiae*, Newton appears to attribute to the "impulse" elicited by way of opposition to impressed force a surprisingly active role in affecting the state of motion of the agent body. Yet *vis inertiae* could not be the ultimate source of material agency. There is a world of difference between "action" and "reaction", between the attraction the sun, for example, exerts upon the earth (when considered *as* attraction, i.e., as rooted in some "active principle" initiating motion) and the reaction of the earth upon the sun (considered *as* reaction, i.e., as rooted in the *vis inertiae* of the earth):

> The *vis inertiae* is a passive principle by which bodies . . . resist as much as they are resisted. By this principle alone there never could have been any motion in the world. Some other principle was necessary for putting bodies into motion, and now that they are in motion, some other principle is necessary for conserving the motion.[85]

Otherwise, all motion would rapidly come to an end, he notes, for it is "always upon the decay" in inelastic or only partially elastic impacts, in motion through viscous media, and so forth.

> Seeing, therefore, the variety of motion which we find in the world is always decreasing, there is a necessity of conserving and recruiting it by active principles, such as are the cause of gravity, by which planets and comets keep their motions in their orbs and bodies acquire great motion in falling, and the cause of fermentation.[86]

To these latter he attributes, in a general way, physiological, chemical, and geological activities, as well as the emission of heat from the sun. So there can be "very little motion in the world" which does not proceed from these active principles, which are somehow extrinsic to matter itself.[87] He continues:

> If it were not for these principles, the bodies of the earth, planets, comets, Sun, and all things in them would grow cold and freeze . . . life would cease and planets would not remain in their orbs.

The conviction that matter of itself is ineffectual to sustain the motions of the world remained firm in the decades after the appearance of the *Principia*, as Newton investigated a wide variety of chemical and electrical phenomena. But he never could decide on just how an "active principle" relates to the matter it operates upon:

> And what that principle is, and by means of [what] laws it acts on matter, is a mystery; or how it stands related to matter is . . . difficult to explain.[88]

We shall return to this in Chapter 3 below. In the meantime, let us note how his uncertainty about the status of these principles and their relations to the forces he wished to quantify led him to avoid all reference to them in the *Principia*. Even the forces themselves, as postulated entities, seemed to have a dubious status from the point of view of the rigorous experimental-deductive method he wished the *Principia* to exemplify, and so he focused rather on the laws of motion which were their observable effects and in terms of which alone forces could be known. In Definition VIII, as we have seen, he says his aim is "only to give a mathematical notion" of central forces. Yet, in the same paragraph, he describes motive force as "an endeavor and a propensity of the whole towards a center", and accelerative force as "a power diffused from the center to all places around so as to move the bodies that are in them", and goes on to speculate on the possibility that the central body might be the cause of the propagation of the central forces, just as a magnet is of magnetic forces. His attempt to restrict the *Principia* to "mathematical notions" evidently could not be fully successful. Here again, one is led to remark that the mathematical network was no more than the skeleton of an explanatory scheme of such physical concepts as *power, propensity, cause, force.*

§2.5   The Impulse to the *Principia*

Yet putting it this way may conceal a difficult and much debated issue. Ought the construction of the *Principia* be regarded as primarily a *mathematical* exercise? Or was Newton attempting to articulate a set of loosely related and rather vague physical concepts by constructing the appropriate mathematical framework around them? What guided him in fashioning the Definitions and Axioms? What role did regulative principles (deriving from metaphysics or theology) or physical intuitions play in the actual elaboration of the *Principia?* This is a very large issue, and we can barely touch on it here. There can be no question about Newton's preoccupation with the ontologies of active principle, force, spirit, and the like, both before and after the *Principia* appeared.[89] But to what extent did the original writing of the work depend on considerations of that kind?

In 1679, Hooke posed Newton the problem: Can one explain elliptical planetary orbits by supposing a "central attractive power" that operates from a focus of the ellipse according to an inverse-square law? It was, as he posed it, a matter of combining two "motions", one along the tangent, the other toward the focus; the difficulty was primarily mathematical because of the need for a method of indivisibles to resolve the question. But he was confident that "finding out the properties of a curve made by two such principles will be of great concern to mankind".[90] How right he was! It was to be another five years before Newton, in the short *De Motu Corporum*, worked out not only an answer to this problem but to several connected central-force problems, thus laying the groundwork for the first section of the *Principia*. The work was almost entirely mathematical in character. The notion of force remained unspecific in it, although, by accepting Hooke's formulation of the problem in terms of a central attractive power, Newton could be said to have committed himself against the older Cartesian mechanical philosophy. The notion of mass was not required, nor was Law III, and yet Kepler's Laws could be successfully derived. This was to remain the most impressively verifiable part of his entire mechanics.

A revision and gradual enlargement of the work followed, with the familiar Definitions and Laws making their first appearance about a year later (1685). Corollary I to the Laws introduces the parallelogram of "forces", as we have seen; Law II is appealed to so as to support an intuitive notion of composition (i.e., the two impulses that act upon the body do not interfere with one another). On this basis, a long procession of theorems follows about orbits under central forces that are assumed to proceed from "immovable centers" (i.e., the effect on the attracting body is neglected). All that is required is that "change of motion" be proportional to the "force" impressed (understood as a sequence of momentary impulses, acting at equal intervals of time). This was the indispensable physical insight which made its appearance here for the first time. This material formed the basis of the first part of the *Principia* when it appeared in 1687.

It was in Section XI and following of the *Principia* that Law III and the notion of quantity of matter were first called upon. He opens this section on centripetal force by noting that although such forces are described as attractions,

> they may more truly be called impulses. But these propositions are to be considered as purely mathematical; and therefore laying aside all physical considerations, I make use of a familiar way of speaking, to make myself the more easily understood by a mathematical reader.

Now while theorems involving such terms as 'impulse' and 'force' cannot be taken to be *purely* mathematical because of the implicit limitations imposed by physical intuition on such concepts, it seems to be the case that the construction of the theorems themselves (especially those in the crucial first three-quarters of Book I, dealing with planetary orbits) was only minimally guided by properly physical considerations. What was mainly (though not exclusively) expounded here was a mathematical method of computing orbits on the basis of an inverse-square law of "attraction". When Newton implied that he had not yet got to grips with the physical implications of the terms he was using, one is inclined to take him at his word. It is clear that the Definitions and Axioms, as we know them, came *after* the successful solution of Hooke's problem. One can regard them, then, as the first step toward a necessary clarification of the physical intuitions embodied in a conceptual system that had been, in the first instance, largely mathematical in its elaboration.

This is not meant to suggest, of course, that Newton's concern with issues of ontology and explanation began with the publication of the *Principia*. His decision not to treat in the *Principia* itself the philosophical issues raised by the notion of force ought not be interpreted as meaning that he had not reflected upon them or that they had played no part in the shaping of the work, but rather that discussion of them at that point would inevitably have alienated the "philosophers" and was unnecessary for the "mathematicians" for whom the work was intended.

In this connection, it is of interest to examine the unfinished *Conclusio* Newton drafted for *Principia* I and left unpublished. He begins by saying that he has now explained the "greater motions" of the universe, the ones that can more easily by detected.

> There are, however, innumerable other local motions which on account of the minuteness of the moving particles cannot be detected, such as the motions of the particles in hot bodies, in fermenting bodies, in putrescent bodies, in growing bodies, in the organs of sensation and so forth. If anyone shall have the good fortune to discover all these, I might almost say that he will have laid bare the whole nature of bodies so far as the mechanical causes are concerned. I have least of all undertaken the improvement of this part of philosophy. I may say briefly, however, that nature is exceedingly simple and conformable to herself. Whatever reasoning holds for greater motions should hold for lesser ones as well. The former depend upon the greater attractive force of larger bodies, and I suspect that the latter depend upon the lesser forces, as yet unobserved, of insensible particles.[91]

The direction of the argument seems clear. The sort of reasoning that worked for the planets ought also to work for the minute particles on which chemical processes depend. Though his own work had not been primarily concerned with such processes, he is confident that the notion of attraction will be helpful in explaining them. In this way, the "mechanical causes" are to be discovered. (In a passage that was dropped from the fair copy he had made of the *Conclusio*, he is more cautious, noting that he does not wish to define how attraction operates. All forces are "attractive" by which bodies are impelled to each other, "whatever the causes be".) The last sentence of the cited passage is significant: the motions of greater bodies are said, without qualification, to depend on the operation of forces, that is, this is taken to have been proved, whereas the operation of forces to explain chemical process and the like is as yet only suspected.

The passage continues:

> For, from the forces of gravity, of magnetism, and of electricity, it is manifest that there are various kinds of natural forces, and that there are still more kinds is not to be rashly denied. It is very well known that greater bodies act mutually upon each other by those forces, and I do not clearly see why lesser ones should not act on one another by similar forces.

Thus the operation of gravity is likened to that of electricity and magnetism; all three are "well known" to require forces by which bodies act upon one another mutually. That similar forces should operate in the invisible realm can be postulated by analogy. When processes involving acids and alkalis occur, there is a violent motion of the minute particles. What is more likely than that such motions are due to forces operating between the particles? It is not, therefore, that he is validating the notion of attraction in mechanics by its prior acceptability in chemistry but rather the other way round. The function of the chemical instances, as of the others, is to underline the potential explanatory fruitfulness of the concept of attraction, not to suggest that active principles are already known to be at work in chemical process; "so why not admit their operation in gravitational motion?" This is not to say that Newton was not made more receptive to the notion of attraction in constructing his mechanics by his early familiarity with the active principles of neo-Platonism and of the alchemists. However, this is a long way from claiming that the acceptability of the attractive-force model in mechanics derived primarily for him from a conviction that it already could be seen to work in the domain of living or of chemical process.

One can see why Newton would be likely to have second thoughts about the wisdom of publishing this argument as the final conclusion of the *Principia*. Its argument for the propriety of the notion of attraction is cogent, resting on the wide variety of phenomena it can account for. Besides magnetic and electrical phenomena, the most obvious cases, there are the coherence of bodies, capillary rise, the penetrability of certain substances by certain other substances, the formation of crystals, the impossibility of perfect contact between bodies, and so on. All of these can be explained by postulating forces between the minute particles that are postulated to make up visible bodies. But this double postulation, and this explicit defense of attraction, were not going to recommend his book to the "mathematicians" for whom it was in the first instance intended. Better to play down all hypothetical elements of this sort and rely on the plain virtues of his mathematical formulation.

§2.6 Force in Matter?

The dilemma about the nature of force, in which he found himself, derived from his espousal of forces as the real causes of new motion (against Descartes), while denying them any inherent status in matter (against Leibniz).[92] When Cotes was writing his Preface for the second edition, he was understandably puzzled by Newton's claim that his use of such words as 'attraction' was not intended to attribute activity to matter. In a letter to Newton, he argued that the axiom "all attraction is mutual" would not be true unless the word 'attract' is used in its proper sense, and he added, rather tartly, that "until this objection be cleared, I would not undertake to answer anyone who should assert that you do *hypothesim fingere*". Newton's answer is curious; he says that mutual attraction is simply an instance of the action-reaction described in the Third Law, and that this Law, far from being a hypothesis, is "deduced from phenomena and made general by induction."[93] And at this point in the new edition he added a further corollary on mutual attraction, in which Jupiter and Saturn are said to "disturb each other's motions".[94] Not only does he not meet Cotes' objection, but in a way he strengthens it. He could still consistently say that God moves *both* bodies so as to simulate an action-reaction pair; in this way the passivity of matter might be retained.[95] This is the sort of "occasionalist" solution that tempted him in the 1690s (as we shall see in chapter 4 below) and that Berkeley would argue for later. However, it was incompatible with the causal language of the *Principia*, and would constitute abandonment of physical explanation in mechanics—surely a high price to pay.

Newton's mode of using the concept of *quantitas materiae* makes it impossible for him, as we have seen, to accept the total indifference of matter to motion. He sees matter as resisting change of motion, even exercising a "force" to prevent such change. In this limited sense, matter is unquestionably active for him. But in the stronger sense of *initiating* motion as the source of impressed forces, he can never quite accept the plain implications of the language he uses, and will continue to insist that "bodies cannot move themselves",[96] leaving himself the thankless task of finding a home for forces somewhere else than in matter.

Why *did* he take this line? His uneasiness about the idea of action at a distance provides one answer, as we shall see later, but there were also other reasons why he tended to see matter as a resistant (though not wholly passive) element only.

§2.7   Matter and Spirit

The first reason was the matter–spirit dichotomy, so dominant in the neo-Platonic and the alchemical traditions, both very influential in shaping Newton's thought, as we have seen. "Spirit" was characterized as the cause of all movement. The motivation for this can be clearly seen in a draft for Query 31(23):

> We find in ourselves a power of moving our bodies by our thoughts . . . and see the same power in other living creatures but how this is done and by what laws we do not know. We cannot say that all nature is not alive.[97]

He thought of an entity which originates activity as "spirit", even when, on other grounds, it might be regarded as "material". For example, he alludes to

> an electric spirit which reaches not to any sensible distance from the particles unless agitated by friction. . . . And the friction may rarefy the spirit, not of all the particles in the electric body, but of those which are on the outside of it.[98]

A "spirit" which has variable extension and variable density and can be "agitated" by friction would surely have been called "matter", rather than "spirit", were it not for its ability to initiate motion. 'Matter' was for him, first and foremost, a contrasting term with 'spirit', the latter being understood rather broadly to include not only God but also the natural agencies responsible for the "violent" motions of chemical and electrical action and even, perhaps, for accelerated motion in general.[99] The contrast was thus not the Cartesian one; 'spirit' meant something much narrower for Descartes. And it was not always a sharp contrast either. Newton represents the spirit–matter relation at times as a sort of spectrum. For instance, the vapors that "arise from the sun, the fixed stars and the tails of comets" have a high degree of causal efficacy;[100] they may, he thinks, provide the life-sustaining "spirit" which is "the most subtle and useful part of our air".[101] Yet, on the other hand, they are also attracted downward to the earth (and in that sense are "material", possessing gravity); they are gradually transformed into partially active corporeal agencies (such as salts and sulphurs) and finally into totally passive things (mud and clay). Clearly, he is thinking of a continuum of possible blends in which the level of activity corresponds to the degree of "spirit".[102]

The second consideration which may have influenced Newton in his decision to make matter as inert a principle as a consistent mechanics would allow was specifically theological in its inspiration. He believed the Christian doctrine of Creation to imply the total dependence of the world on God's activity, and he often tended to interpret this to mean that the activity in the world had to come directly from God, without any secondary intermediary. To locate the active principles responsible for motion in matter, as Leibniz did, was to make matter, once created, a self-sufficient entity.[103] To Newton, this seemed tantamount to atheism; he was still as intent as Aristotle had been to find a First Mover at the summit of his mechanical system.[104] It was, in his view, quite improper to suppose that the motions of physical bodies could be explained without recourse of any kind to the power of God.

Ironically, Leibniz' ground for attacking this view was equally theological.[105] He thought it "a very mean notion of the wisdom and power of God" to suppose that constant divine interventions are necessary to keep the universe operating properly; to take matter to be totally inert, he argued, would unduly limit the creative options open to God.[106] His stress was thus on the limits set on God's absolute power by His (free) acceptance of the demands of intelligibility in His creation. Like Descartes, he made God's relation to the universe such that a rationalist metaphysics would be possible, one which began from the assumption that the universe is open to the human mind by way of the Principle of Sufficient Reason. Newton's natural theology, on the other hand, was much more squarely in the voluntarist tradition: since God is free and all-powerful, we cannot know *a priori* what laws we will find in nature, and thus they have to be determined inductively.[107]

CHAPTER THREE

Is Gravity an Essential Property of Matter?

THERE WERE, as we have just seen, two main obstacles to Newton's holding matter to be of itself inert, as he evidently wished to do: his attribution of *vis inertiae* to it and his postulation of gravitational attraction between bodies. Of the two, the latter is clearly the more serious: if every body attracts every other (as the *Principia* clearly implies),[108] how can matter be thought of as passive? Indeed, does not gravity (understood as a power of acting on other bodies) become an *essential* property of matter?

§3.1 Absolutely Not!

Newton's response to this question throughout his career was a definite no. Before we ask what he meant, it is important to appreciate why he was so emphatic in this denial. He notes two reasons.

> You sometimes speak of gravity as essential and inherent to matter. Pray, do not ascribe that notion to me; for the cause of gravity is what I do not pretend to know and therefore would take more time to consider of it.[109]

This was in a letter to Bentley (1693). In 1717 he was still unsure, and emphasized this by putting the following note in an "Advertisement" on the front page of the second edition of the *Opticks:*

> And to show that I do not take gravity for an essential property of bodies, I have added one question concerning its cause, choosing to propose it by way of a question, because I am not yet satisfied about it for want of experiments.[110]

Thus his first reason for not wanting to portray gravity as essential to matter was that such a claim would wrongly suggest that he understood its working. By 'understand' Newton still meant what his critics meant: 'understand in mechanical terms of contact action'. The generally accepted doctrine of primary and secondary qualities was based on the assumption that the primary qualities should need no further explication; they should, by preference, be immediately given in sensation. From this perspective, gravity (power to attract) sounded uncomfortably like the occult quality that Huyghens and Leibniz claimed it to be. Thus the "cause of gravity", that is, the cause of bodies approaching one another as though there were in them a power of attraction, would still have to be sought; it would not suffice, Newton was persuaded by his critics, to postulate gravity as essential to matter, in need of no further explication. Such a postulate might even be taken by some to suggest that he could somehow exclude explanations in terms of hypothetical intermediaries like aether or spirit, something he was very far from holding he could do.

But there was an even more basic reason, Newton more than once made clear,[111] as in another letter to Bentley (1693):

> It is inconceivable that inanimate brute matter should, without the mediation of something else which is not material, operate upon and affect other matter without mutual contact, as it must be, if gravitation . . . be essential and inherent in it. And this is one reason why I desired you would not ascribe innate gravity to me. That gravity should be innate, inherent, and essential to matter, so that one body may act upon another at a distance in a vacuum

without the mediation of anything else by and through which their action and force may be conveyed from one to another, is to me so great an absurdity that I believe no man, who has in philosophical matters a competent faculty of thinking, can ever fall into it.[112]

In this well-known passage, Newton links the notions of innate gravity and action at a distance, and concludes that since the latter is unacceptable, the former must be rejected too. To make matter "essentially" attractive would suggest that no other intermediary is needed in order for the attraction to operate; in that event, even if two bodies were separated by a vacuum they would still cause one another to move. This was the consequence that Newton's contemporaries had found so shocking, and which Newton was anxious to disavow, though whether he ever really thought it as great an absurdity as he here suggests is by no means certain.[113] But there can be no doubt that his unwillingness to attribute gravity to matter as an essential property derived in part from this source. This of itself would not, of course, entail the view that matter is totally inert (as the example of Leibniz showed). However, it would seem to undercut the strongest argument for attributing activity to matter in a Newtonian universe.[114]

§3.2 "Gravity"

What made this whole question even more intractable was the ambiguity of Newton's terms, 'gravity' and 'essential'. It will be helpful to point up some distinctions that Newton did not make explicit, in an attempt to unravel the multiple meanings that the term 'gravity' was later seen to possess. The gravity of a body can be construed either as active (A) or as passive (P), depending on whether the body is considered as attracting or being acted upon. It can be actual (E) or dispositional (D), depending on whether gravity is attributed to a body on the basis of its actually exercising attraction (or actually being attracted) or only on the basis of its capacity to act (or be acted upon). Thus, for instance, a body alone in a void would not exhibit gravity in the former sense, but would in the latter. Finally, 'gravity' could refer either to the weight (W) or to the mass (M) of a body.

In Definition V of the *Principia*, where the term 'gravity' is first introduced, Newton speaks of it both as a "force" and as "that by which bodies tend to the center of the earth". A projectile is drawn out of its rectilinear path "by its gravity" or "by the force of its gravity". The metaphors of agency are plain: gravity is not just a neutral tendency to move in the presence of other bodies. It is a force which is elicited in a body and alters its state of motion. Or it is its mass: "the less [the projectile's] gravity is, or the quantity of its matter...the less it will deviate". It was not necessary for Newton to distinguish here between mass and weight; 'gravity' evidently covered both.

But this would no longer do when the question Is gravity essential to matter? was raised. It then became crucial to know which sense of gravity was intended. E- and W-gravity could not qualify as essential, since a body alone in a void would have neither. PD-gravity, the disposition to be acted upon by other bodies (or the tendency to move in the presence of other bodies), would presumably be a universal quality. The troublesome one, from Newton's standpoint, was AD-gravity (the disposition to attract other bodies), with its overtones of action at a distance. To regard it as essential (or even as universal) would leave him open to the sort of objection to which he was most sensitive, as the letter to Bentley illustrates.

But could they be separated? Could one sort of "gravity" be accepted as universal without others being carried along? Could one admit the universality of "heaviness towards the earth" among terrestrial bodies without also postulating an equivalent universality for the active property of attraction apparently displayed by the earth? Newton could not, of course, have posed the question in this way since the requisite distinctions had not yet been drawn.[115] But even if he could have, the answer would not have been apparent because of his uncertainty as to how and where "active principles" should be located.

§3.3 "Essential" to Matter

A second set of difficulties is associated, as we have seen, with the term 'essential'. Newton sometimes used it in a loose sense, as in the letter to Bentley quoted above, where he equates it with 'innate' and 'inherent in', two very different terms in the technical usage of that day. The closest he came to formulating a criterion for its use was in the puzzling intensity-invariance language of Rule III, though he was reluctant even then to tie himself too firmly to the metaphysical-sounding term 'essential', rather than the more neutral 'universal'. There seemed to be no doubt that gravity *did* satisfy this criterion. In *Principia* I, he wrote of the forces acting on the tails of comets:

> As the gravity of terrestrial bodies is proportional to the matter in them, therefore, if the quantity of matter remain the same, the gravity cannot be intensified or diminished.[116]

In a draft revision of Query 31 (c. 1716), he wrote:

> All bodies here below are . . . heavy toward the earth in proportion to the quantity of matter in each of them. Their gravity in proportion to their matter is not intensified or diminished in the same region of the earth by any variety of force and therefore cannot be taken away.[117]

There is, of course, the familiar ambiguity as to which sort of "gravity" *is* intensity-invariant. In the first passage above, it

sounds like mass; weight could be altered by changing the
position of the body. In the second, it sounds like weight, but
it is the gravity/mass ratio which is said to be intensity-
invariant.[118] Clearly, mass lends itself best to the intensity-
invariance sort of claim. Clearly, also, Newton took gravity
(perhaps in more than one of its several senses) to be intensity-
invariant.[119] Was it, then, a universal quality, like mobility?
The answer would seem obvious, but so great was his reluc-
tance to allow quasi-essential status to gravity that he had to
make sure that no one would attribute to him this further
unwanted conclusion.

§3.4   Universal Gravitation

The entire argument of Book III of the *Principia* ("The
System of the World") depended on the universal applicability
of the mechanical principles of Book I and, specifically, on the
uniformity of the weight/mass proportionality, so that any two
bodies, placed in the same position, would "gravitate" (accel-
erate) to the same degree. The first propositions of Book III are
concerned with this issue. Their argument is that the proportionality of weight and mass, demonstrated for terrestrial
bodies by the pendulum experiments and for the planets and
planetary satellites by an analysis of their orbital motions, can
be safely generalized for *all* bodies. For someone who emphasized inductive method as much as Newton did, this sort of
generalization required some sort of justification. When one
says that every particle of the earth attracts every other particle,
one must immediately concede that "no such gravitation anywhere appears", the reason presumably being that this effect
"must be far less than to fall under the observation of our
senses".[120]

Corollary II to Proposition 6 addressed the matter of
justification. An interesting shift occurred in the text of the

Corollary between the first and second editions. In the first edition, after repeating the inductively based assertion of Proposition 6 that "all terrestrial bodies are heavy towards [gravitate towards] the earth, and the weights of all that are at equal distances from the center are as their quantities of matter", he went on to argue that if any body (the aether included) were to be devoid of gravity or even were to "gravitate less in proportion to its quantity of matter", this would lead to contradiction, since, according to the principle of universal intertransmutability (Hypothesis III of this edition), any such body could be transmuted by degrees into bodies with normal gravitational properties. This would make weight depend on the form of the body, "altogether against experience", as the previous Corollary had asserted.

The argument appears to be circular, since if it is already known that weight does not depend on the form of the body, there cannot be bodies devoid of gravity or with a different weight/mass proportionality—the conclusion the argument is supposed to validate. In his annotated copy of the first edition, Newton crossed it out, replacing it with a simple reference to Hypothesis III: "if it should hold here". It was not, however, deleted in later editions, but its logical status was entirely altered in the second and following editions by the insertion of the following sentence before it, immediately after the assertion about the gravitational properties of terrestrial bodies: "This is the quality of all bodies within the reach of experiment, and therefore by Rule III is to be affirmed of all bodies whatsoever". Thus the status of gravity as a universal quality of all bodies is now asserted on the strength of Rule III—and Rule III apparently shorn of the intensity-invariance criterion.[121] Gravity can be regarded as a quality of all bodies simply because it has been found to be a quality of all those bodies that are within the range of observation and experiment, nothing more. It is thus on the same footing as the other "universal qualities" listed with it in Rule III.[122]

Though it might seem from the inductivist standpoint that gravity was a much less obvious candidate for universal status than the other properties mentioned in Rule III, Newton was apparently determined not to allow any distinction in this regard between them:

> If it universally appears, by experiments and astronomical observations, that all bodies about the earth gravitate towards the earth, and that in proportion to the matter they severally contain... that our sea gravitates towards the moon and all the planets towards one another... we must in consequence of this rule, universally allow that all bodies whatsoever are endowed with a principle of mutual gravitation. For the argument from the appearances concludes with more force for the universal gravitation of all bodies than for their impenetrability of which, among those in the celestial regions, we have no experiments nor any manner of observation.[123]

Since impenetrability had already been declared to be a universal quality, this was about as affirmative a statement as one could imagine.

What is meant here by the "universal gravitation of all bodies"? A single capacious term covered a cluster of concepts, as we have seen: what is universal in the property is that all bodies have to be acted upon (P-gravity) in proportion to their mass (M-gravity) and thus exhibit the phenomena of weight (W-gravity). The troublesome sense is the active one (A-gravity), with its overtones of action at a distance. Yet in this part of the *Principia* the active character of gravity is stressed. In Proposition 5, the force of gravity is said to "pull" planets from rectilinear motion to elliptical paths; the planets also "gravitate to [are heavy towards] one another", to the point that Jupiter and Saturn "by their mutual attractions appreciably disturb each other's motions".[124] To be "heavy towards" thus implies an active influence of each upon the other. In the third edition, after this Proposition, he added a brief Scholium:

The force which retains the celestial bodies in their orbits has been hitherto called centripetal force; but it being now made plain that it can be no other than a gravitating force, we shall hereafter call it gravity. For the cause of that centripetal force which retains the moon in its orbit will extend itself to all the planets, by Rules I, II and IV.[125]

Rules I and II (which deal with causes) are invoked here because it is the *cause* of gravity that he is universalizing. In the General Scholium, he notes that though the phenomena of the heavens and the sea have been explained in terms of gravitational force, the *cause* of this force has not yet been determined, even though its universal extent is known:

> This is certain, that it must proceed from a cause that penetrates to the very centers of the sun and planets without suffering the least diminution of its force... and propagates its virtue on all sides to immense distances.[126]

But this is not enough to specify what the "cause of the properties of gravity" is; to suggest a cause would require one to go beyond the phenomena, that is, to "prove an hypothesis". It is at this point that he brings forth his celebrated injunction against hypotheses ("whatever is not deduced from phenomena"), which are to be excluded from "experimental philosophy". Instead, particular propositions are to be "inferred from the phenomena and afterwards made general by induction". It was by means of such a method, he avers, that the impenetrability of bodies and the laws of motion and of gravitation were discovered:

> And to us it is enough that gravity does really exist and act according to the laws which we have explained and abundantly serves to account for all the motions of the celestial bodies and of our sea.[127]

Newton's dilemma was nowhere made clearer than in these pages. If he insists on deductivism, gravity can be no more than a tendency on the part of a body to move toward other bodies according to the inverse-square law, interpreted descriptively. But once he starts talking about planets disturbing other planets' motions, of a force which retains the moon in its orbit, or even of gravity's "accounting for planetary motions", he has moved from the level of description and generalization to that of explanation, and, inevitably, hypothesis. In the General Scholium, he attempts to introduce a second level of explanation, a mysterious "cause of gravity" which must "penetrate to the very centers of the planets", and which differs in obvious ways from the "mechanical causes" of impact mechanics. It is *this* "cause of the force of gravity" that he allows to be hypothetical and thus beyond the reach of experimental philosophy. But since the force of gravity is described over and over again as a cause, it is redundant to introduce a cause of a cause: the embargo on "hypotheses" already affects the notion of gravity itself, once one goes beyond the descriptive language of mass and acceleration to speak of it as a "force".

What is at issue in this passage are not only the limits of inductive method but also the sense in which one can count gravity as a universal quality on a par with impenetrability. For Rule III to apply to it, it must be "found to be a property of all bodies within the reach of experiments." But to what extent *can* he say that gravity *has* been found to be a property of bodies? In the sense of "tendency towards", yes; in the sense of mass, yes. But in the sense of active powers of attraction or of centers of force, no. Gravity, in these senses, is an explanatory hypothesis, a postulated cause; the "law" of gravitation is a *theory* of gravitation, a well-supported one, no doubt, but assuredly not something that could be deduced from the phenomena, as Berkeley would be quick to point out.

What prevented Newton's position from being open to an immediate charge of inconsistency was, of course, the ambiguity of the term 'gravity', as he used it. In a draft of *Opticks* III (1716) he wrote:

> I reckon the [primordial particles] to be hard and heavy as well as impenetrable, because I meet with equal evidence for all three qualities. For we have no evidence for their being impenetrable but experience in all the bodies we are able to examine, and we have the same evidence for their being hard and heavy. . . . For we have the whole course of a large experience for the universal gravity of matter and for the hardness of its particles without any instance to the contrary, and we have nothing more for its universal impenetrability.[128]

But there is a leap of creative imagination between this sort of "heaviness" and the "heaviness" described in the early propositions of Book III, a leap that bridges a gap as wide as that between Aristotle and Newton. And this, for a good inductivist, should have been enough to eliminate heaviness, in this ontologically stronger and epistemologically risky sense, from consideration as a "universal quality", certifiable on the basis of Rule III. The primary–secondary distinction was based, as we have seen, on the assumption that the primary qualities are immediately given in sensation; in order to furnish a basis for a mathematized mechanics, they had (so it seemed) to be themselves epistemologically unproblematic. Hypothetical constructs would not do. Impenetrability was acceptable. But gravity?

Newton's last word on the topic was a perturbed (and confusing) disclaimer, added at the end of Rule III in the third edition of the *Principia*: "Not that I affirm gravity to be essential to bodies. By their *vis insita* I mean nothing but their inertia. This is immutable. Their gravity is diminished as they recede from the earth". He had just asserted that mutual gravity is a *universal* quality of all bodies. But he now goes on to add that this ought not be taken to imply that it is "essential" also. As we saw above, he had constantly tried to disavow this implication. In his letters to Bentley, he had rejected it because it would have held him (he thought) to explaining how gravity operates (something "I do not pretend to know") or, even worse, to assuming that bodies could act at a distance.[129] What makes this postscript difficult to reconcile with the paragraphs that precede it is his further remark that gravity diminishes with distance from the earth; that is, it is not intensity-invariant, as Rule III would require a quality to be in order to qualify as universal. Are we to infer from this that gravity is *not* universal, even though the entire argument to this point had apparently demonstrated that it is? And why is the intensity-invariance criterion suddenly re-introduced here? As we have seen, it nowhere figures in the applications of Rule III up to this point. Why, furthermore, is the term 'essential' used, when it nowhere appears in the earlier text? One gets the impression that Newton is trying to avoid at all costs any suggestion that his use of Rule III to draw up the list of universal qualities could be construed as an implicit claim that gravity is an essential property of matter. But by implying that gravity does not satisfy Rule III, on the grounds of its being mutable, he leaves the reader wondering how to reconcile conflicting claims as to whether gravity is a universal quality and whether it is intensity-invariant.

The reference to *vis insita* is perhaps the most puzzling feature of the postscript. The term had nowhere appeared in the preceding discussion, but now he appears to suggest that in talking about the universal gravitation of bodies, he had their *vis insita* in mind, which would qualify as universal (he suggests) being immutable. But this will not do. First, *vis insita*, as we have seen, is measured either by quantity of motion or by impressed force. In neither case is it invariant. Second, *vis insita* is in no way equivalent to the "heaviness towards" of which he had been speaking. But perhaps he is not equating them but merely saying that the inertia of bodies (unlike their gravity) is universal. He had already classified inertia as a universal quality, under Rule III. But this "inertia", if it is to be immutable (as he says it is), cannot be the *vis inertiae*. Since each body is "endowed with its proper inertia", it sounds like a resistance correlated directly with quantity of matter. The most favorable interpretation of the postscript (one that admittedly stretches likelihood) is that Newton is distinguishing mass and weight and belatedly noting that, of the two, only the former could qualify as a universal quality.

§3.5 "Universal" versus "Essential"

Cotes was more forthright about the entire issue. In his preface to *Principia* II he emphasized that the arguments for the universality of such properties as extension were no stronger than (nor in any significant manner different from) the arguments for universality of gravity: "in short, either gravity must have a place among the primary qualities of all bodies, or extension, mobility, and impenetrability must not".[130] He had. indeed, first written 'essential' here, but changed it to 'primary' at Clarke's suggestion so as not to "furnish matter for cavilling".[131]

One can scarcely blame Cotes for being confused by Newton's hedging.[132] They could hardly have failed to discuss the text of the new Rule, since it was one of the main changes in the new edition in Cotes' charge. Cotes was willing to drop the word 'essential', not because he attributed a more problematic status to gravity than to other properties but (as he makes quite clear in the letter to Clarke already quoted) because he was skeptical whether *any* properties, extension included, could be shown to be "essential" in *his* sense of the term: an essential property is one "without which no other belonging to the same substance can exist". He insists that insofar as we can talk of "essential" properties at all, gravity has "as fair a claim to that title" as properties like extension do. This is surely a correct reading of Rule III, as it appeared in the second edition. Newton, by his later emendation of the Rule, suggested a distinction between 'essential' and 'universal' (or 'primary'), thus, in effect, adding an extra criterion for essentiality, namely, that the property involve no hypothetical aspect of "hidden" agency. In this way he hoped to evade the charge of condoning action at a distance, which clearly went beyond the "given" of the inverse-square law to the assignment of hypothetical agency.

But now the defect in his strategy is clear. To allow gravity to be a *universal* quality would incur exactly the same risk as would the rejected doctrine of the "essentiality" of gravity: requiring action at a distance. What would have been needed in 1726 was a thorough revision of the Rule; but by this time, Newton may not have wanted to reopen a problem that had from the beginning proved so intractable. It would have been necessary to clarify the implicit distinction between two senses of 'gravity', instead of relying on a distinction between 'universal' and 'essential'. One of these *would* be universally predictable of matter, and the other would not. Had he not, for instance, suggested a proof of sorts for the claim that no piece of matter can be wholly devoid of "gravity?"[133] To say that "gravity" is essential to matter in *this* sense was to do no more than deny the possibility of bodies which would exhibit none of the phenomena of weight. Leibniz' claims that there could be "effluvia" (such as light) which had none of the tendencies to falling or orbital motion of ordinary bodies was taken by Clarke to imply the doctrine that "gravity is not essential to matter", in the sense that there are pieces of matter that lack weight. Clarke opposed this, and equivalently argued that such effluvia ought not count as "material".[134] To him (as to Newton), it was obvious that the phenomena of weight were invariably associated with bodies; it was upon this supposition that the *Principia* had been based. But this was a far cry from claiming an understanding of how gravity operates.

Such a distincition was implicit, as we have seen, in the frequent attempts Newton made in the *Principia* to restrict that work to an ontologically neutral sense of such dynamic terms as 'force', 'attraction', 'gravity'—a sense which would be descriptive and mathematical, merely attributing certain sorts of regularity of motion to one body in proximity to another, without postulating the kind of agency responsible. M-gravity, in this sense, would be neither active nor passive; to call it a universal property would imply only that every body has a particular way of behaving in the presence of other bodies, and that this property is intensity-invariant. It would not have to be regarded specifically as a response to the activity of the other bodies or to forces considered as real extrinsic agencies of a well-defined kind.

One trouble with this, of course, was that Newton's terminology was difficult to de-ontologize in this way. When he spoke in the *Principia* of bodies "attracting", or "pulling", or "acting upon" one another, it was difficult to take these words as figures of speech, connoting nothing whatever in the way of agency, either active or passive, on the part of the bodies involved.[135] Not only were his critics unpersuaded, but Newton, in his incessant drafting and redrafting, was just as prone as they to take such terms as 'attraction' to mean what they say.

§3.6 Gravity and the Mechanical Philosophy

More serious, though, was Leibniz' objection that if one prescinds from all questions of agency and remains at the purely descriptive level, one cannot properly be said to have *explained* motion at all. He caustically remarked that his English rival had to fall back either on occult qualities or on miracles, the only alternatives he could see to the form of mechanical action than he himself favored.[136] From the beginning, Newton knew very well that the concept of gravity formalized in the *Principia* could not be interpreted in "mechanical" terms, as the label 'mechanical' had been understood up to that time. He had proved (as Cotes later emphasized) that the interplanetary spaces were void of all (or virtually all) resisting matter, so that there was no medium across which contact action, of the sort demanded by the mechanical philosophy, could be established. Besides, gravity penetrates to the center of the planets (as he notes) in a way which is dependent of their surface areas, which is not how "mechanical causes" operate.[137] Later, as we shall see in a moment, he made a last attempt to restore an aether which would be the source of gravitational motion, but, significantly, this aether would be distinctly non-mechanical

Newton had no particular reason to want to resurrect the older mechanical philosophy. He had always been open to the diversities of explanatory principle favored in the non-mechanical traditions of alchemy and neo-Platonism. His early conviction, that the "vital motions" associated with chemical, physiological, and such changes could not be reduced to mechanical impact of corpuscles, was to deepen with the years. In response to Leibniz' universal "mechanism", he tended to respond by limiting more and more the scope and value of mechanical explanation, suggesting on one occasion, for example, that the supposedly "mechanical" properties of extension and inertia are, after all, themselves "perfectly incapable of being explained mechanically". So why should *gravity* be singled out as in some objectionable sense "occult" on the grounds that it cannot be handled within the limits of classical mechanical explanation?[138]

To say that an explanation is not mechanical ought not, however, be taken to imply that it is not "natural", as Clarke reminded Leibniz: "The means by which two bodies attract one another may be invisible and intangible and of a different nature from mechanism; and yet acting regularly and constantly may well be called natural".[139] Of course, if someone identifies 'natural' and 'mechanical' (as Leibniz, in effect, does), then all regular motions in the world (including human actions) would become "mechanical" by definition. But now the word has become empty. If, however, it is used in what would have been the ordinary sense at that time, Clarke can quite permissibly say that "gravitation may be effected by regular and natural powers, though they be not mechanical".[140]

Newton's own use of the term 'mechanical' is by no means consistent. Besides the narrow sense the term had for the mechanical philosophers (equivalent to "explicable in terms of contact action"), it also on occasion meant for him no more than "explicable in terms of forces". Thus, for example, in the Preface to the *Principia* he writes: "I wish we could derive the rest of the phenomena of Nature by the same kind of reasoning from mechanical principles, for I am induced by many reasons to suspect that they may all depend upon certain forces".[141]

Gravitational explanation would automatically qualify as "mechanical" in this much broader sense, no matter what agency were ultimately invoked to explain the precise way in which the motion was brought about. But this was certainly not the sense of the term with which his contemporaries were familiar. It is misleading, therefore, to take Newton to be an exponent of the "mechanical philosophy".[142] If this was a mechanical philosophy, it was compatible in his mind with the widest variety of explanations of how gravitational motion actually occurs, including action at a distance and God's direct intervention. Newton himself would hardly have called his views on this topic a "philosophy" (let alone a mechanical one). To use the mathematical-physical notion of a force was not, as he strongly stressed, to commit oneself to any theory of how it operates, to a "philosophy". His program for physics and chemistry—to seek out the laws of motion characteristic of each domain (which would automatically determine what the "forces" were)—was not equivalent to a "philosophy", as that word would then have been understood. It was a proposal as to how experimental research should be guided, and it was open (as he well knew) to different philosophic interpretations. In the next chapter, we shall return to a discussion of some of the interpretations he attempted.

CHAPTER FOUR

How Is Matter Moved?

NEWTON WAS the inheritor of two very different traditions in natural philosophy, and nowhere were their differences more apparent than on the question of how matter is moved. The Cartesians insisted upon contact as the only possible mode of action; followers of the loosely allied neo-Platonic and alchemical traditions admitted the widest variety of "active principles", as did those whose approach to knowledge could be called "hermetic". The Cartesians were guided by the criterion of intelligibility; their temptation, therefore, was to a severe reductionism. The others were inspired by macrocosm–microcosm analogies that suggested organic or human modes of acting as a better guide; *their* temptation was to excessive speculation or elitist mysticism. Newton's laborious and life-long effort to explain the motion of matter can be described as a dialectic between these two quite opposing poles, a dialectic he never succeeded in resolving, but which he carried far beyond the point at which he had found it.

§4.1 Early Solutions

The first section of the *Questiones* begun during his under-graduate days, was headed "Of first matter". In it he opts for atomism. But all action had to be due to contact, and so he spoke of a "matter causing gravity", the invisible aether essential to a mechanical philosophy.[143] On the other hand, in the early *De Gravitatione* tract (dating probably from the later 1660s), the outlines of a quite different model are sketched. The basic insight is that of space as "a disposition of being *qua* being", an absolute within which bodies move (unlike the relative space of Descartes), extensionally identical with God.[144] Bodies are no more than volumes within that space endowed by God with impenetrability and mobility; they are moved by Him much as *our* minds move *our* bodies. In this way he can "show that the analogy between the Divine faculties and ours is greater than has formerly been perceived by philosophers" and "thoroughly confirm" the "chief truths of metaphysics", notably the existence of God. He challenges the reader to say which of the other natural philosophies of the day can be said to do the same:

> If we say with Descartes that extension is body, do we not manifestly offer a path to atheism, both because extension is not created but has existed eternally, and because we have an absolute idea of it without any relationship to God?[145]

His motivation in exploring this ingenious analogy was thus in the first place theological, but one immediate consequence was the attractive alternative it provided to the aether of the mechanical philosophy in handling the intractable problems of communication of motion. It was only twenty years later, with the storm that followed the publication of the *Principia*, that Newton began to explore this model anew, with the added motivation of avoiding the incoherences of the other main alternatives open to him regarding gravitational action.

In the meantime, the *Hypothesis of Light* (1675) introduced yet another possible way out of the impasse brought about by the too-severe reductionism of the mechanical philosophy, but one even more alien to the spirit of that philosophy, perhaps, than God's direct action would have been. Though the basic explanatory metaphor of the *Hypothesis* is that of an aether,

Newton distinguished between "the main body of phlegmatic aether" (which resembles the Cartesian aether in being inert and thus able only to redirect motion) and the "various aethereal spirits" responsible for the *initiation* of motion in refraction, electrical attraction, gravitation, muscular action, and other such phenomena where new motion clearly appears.[146] Such a spirit is akin to a "humid active matter"; the "gravitating attraction of the earth" may thus be attributed to the action of an aethereal spirit which bears "much the same relation to aether which the vital aereal spirit, requisite for the conservation of flame and vital motions, does to air".[147] The permeability of matter to such a spirit may depend not so much on the latter's subtility as on its "sociability"; he notes that various fluids are prevented by "some secret principle of unsociableness" from mixing and, further, that the addition of a third can cause two formerly "unsociable" substances to blend (e.g., tin added to lead and copper). In the same way, "the aethereal animal spirit in a man may be a mediator between the common aether and the muscular juices to make them mix more freely".[148] Though the metaphors he used to describe the operations of "spirit" are quite mechanical (condensation, pressure, squeezing), the important point is that spirit is *active*; it contains within itself the principles of its own activity.

These "active principles" recall the traditions of Renaissance natural magic:

> God, who gave animals self-motion beyond our understanding, is without doubt able to implant other principles of motion in bodies, which we may understand as little. Some would readily grant this may be a spiritual one; yet a mechanical one might be shown, did I not think it better to pass it by.[149]

Analogies drawn from human action were characteristic of the hermetic and the neo-Platonic traditions, while "aethereal spirits" were invoked by the alchemists to explain the same phenomena for which Newton now suggested them. We are a long way here from the canons of the mechanical philosophy. Yet Newton does not want to depart from them further than he can help, as the cryptic note at the end of this last passage indicates. Its contrast between 'spiritual' and 'mechanical' is puzzling, however, since the *Hypothesis* depends so largely on "spirits". And although one might use the hydrodynamic analogies of the aether theorists in describing their operation,[150] there was no way in which a multiplicity of self-active vital principles, governed by laws of sociability, could be reduced to the conceptual straitjacket of mechanical explanation, as the term 'mechanical' would have been used by Newton's contemporaries.

Had Newton's active principles remained at the qualitative and speculative level at which they appear in the *Hypothesis*, however, they would not have accomplished any more for him in mechanics than they had accomplished earlier for van Helmont and the Paracelsians in chemistry. In the next decade, as he tried to chart the endless variety of chemical interactions, both through an intensive program of readings in alchemical texts and through extensive and careful experimental research on his own account, his assurance grew that active principles were everywhere at work in Nature. And gradually he began to associate them with *forces*, now no longer limited to contact action or the operation of density gradients in an aether but operating to attract or repel bodies at a distance from one another.[151] In an unfinished paper, *De aere et aethere* (1679), he argued, on the basis of a rough quantitative law of force, that "some principles acting at a distance" to repel furnished the best explanation of Boyle's Law.[152] But it was, of course, the *Principia* that brought the problem finally to a focus; the inverse-square law of gravitation around which the entire work is built required the postulate of attractive forces operating at a distance. And these forces had to be physically *real*; the reality of (absolute) motion depended upon this.

§4.2   An Ontology for the *Principia*

As Newton reflected upon the agencies involved in gravitation, four broadly different explanatory models would have suggested themselves to him, each having the warrant of a particular philosophical tradition behind it. Most attractive would be an aether, of course, but it obviously could not be the conventional mechanical one; it would have to be of zero, or almost zero, density and operate on matter in some way other than by pressure. A second, less likely alternative would be to seek in light (as the neo-Platonists had always done) a source for all activity, including even gravitational action. A third possibility, suggested by the alchemical tradition, was a localized "spirit" or "active principle", associated with matter in some way, though ultimately distinct from it. Such a principle, not being limited by material modes of causation, could act across spatial gaps; it would be more plausible in explaining short-range forces, perhaps, than in accounting for the long-range forces of gravitation, linking each body with every other. Finally, one could call upon the power of God directly, as neo-Platonists like More had done; could He not move matter in space just as soul moves body? Thus the "active principles" would become identical with God's omnipresent power.

It is fascinating to follow Newton's thought over the next forty years, as he wrestled with the difficulties involved in each of these alternatives and kept moving from one to the other in the endless drafting and redrafting that preceded the Queries of the *Opticks* and the later editions of the *Principia*. Though he stressed different models at different times, one must be wary of imposing too neat a periodization, as though he explored each model in turn in an orderly way. In point of fact, none of these explanatory models is ever repudiated by him; all continue to play a part in his shaping of the ontological alternatives.[153]

In the period of debate that followed the appearance of the

Principia, he often used this sort of metaphor:

> If . . . all space is the sensorium of an immaterial, living, thinking being . . . whose ideas work more powerfully upon fit matter than the imagination of a mother works upon an embryo . . . the laws of motion arising from life or will may be of universal extent.[154]

Because such action followed invariable laws, it could be called "natural" (to counter Leibniz' reproach). And since the laws could in principle be determined empirically, even chemical and physiological changes (due, he suggests once again, to short-range forces of attraction or repulsion between minute particles) could be understood in terms of the categories of the *Principia*. Or could they? What are we to make of the "active principles" of which he speaks so often in these years, or of the analogies he draws with living motion? Do they indicate that there were, for him, two different sorts of motion, one lending itself to analysis in the inertial/material terms of the *Principia* and the other not?[155]

A distinction must be made. Even when Newton speaks in these terms, he is still thinking in universal categories, as in the passage just quoted. Or again:

> And since all matter duly formed is attended with signs of life, . . . if there be a universal life and all space be the sensorium of a thinking being who by immediate presence perceives all things in it, as that which thinks in us perceives their pictures in the brain, the laws of motion arising from life and will may be of universal extent.[156]

But a more fundamental unity is conveyed by the notion of force itself. If this notion can be properly applied in the "invisible realm" (Newton's "analogy of Nature" required one to assume that it could), the *effects* of the forces acting in physiological, chemical, electrical, or other changes would be determined by Newton's laws just as in the case of gravitational motion. The acceleration produced, for example, would be proportional to the force acting on the particle, whether that particle were a planet or a "primordial" corpuscle.[157] If this were not so, it would not be just a question of different laws of force; it would mean that the laws of mechanics would not apply to the microworld at all.

Newton makes his intention clear in this regard. In Query 31(23) he proposes that the various "changes of corporeal things" are to be explained in terms of the motions of the permanent particles of which they are composed.

> These particles have not only a *vis inertiae,* accompanied with such passive laws of motion as naturally result from that force, but also they are moved by certain active principles, such as is that of gravity, and that which causes fermentation and the cohesion of bodies.[158]

Thus the *vis inertiae* of the body acted upon must be taken into account in determining that body's response to an external force, whether that be the force of gravity or the force responsible for fermentation.

But what about the active principles? Are there grounds for distinguishing between the relationship of these principles to matter in gravitational as against other sorts of motion? The passage just quoted seems to set them all on the same level. The first problem lies in the ambiguity of the notion of active principle itself. The passage continues: "These principles I consider... as general laws of nature by which the things themselves are formed, their truth appearing to us by the phenomena, though their cause be not yet discovered".

Here he identifies the "principles" with the laws themselves, made manifest in the phenomena. They are thus fully known, though their "causes" are not. More often, he takes the active principle to be itself the cause, the ontological constituent responsible for the motion. The term is thus for him a loose generic one, which could cover a wide variety of causes, such as will (whether human or divine) and the spirits responsible for fermentation and other chemical changes, as well as gravitational forces. Forces are one kind of *"principium"* or source; only when they are identified with the laws of motion themselves does he seek a causal active principle that is responsible for them in turn.[159] There is no suggestion of a "two-tier" ontology, with forces causing motion and active principles causing forces.[160] Such an ontology would have seemed to his readers perilously close to that of the permanently occult qualities he denounces in Query 31.

But even a one-level ontology of forces, regarded as a species of active principle, was open to question too. Could the forces be described as "incorporeal"? This would put them in the category of substance, and would suggest some sort of immaterial entity operating in the space around bodies. On the other hand, to make them "corporeal" would almost inevitably imply an aether of some sort. Of course, if he did not allow *some* sort of exteriorization of force, the only apparent alternatives were either to postulate forces in matter, operating at a distance, or to deny an ontology to them entirely. One can easily see why none of these possibilities would have seemed especially attractive to him.

Let us return to a comparison of the active principles responsible for gravitational motion, on the one hand, and the various vital motions on the other. Gravitational force is specified by the matter-component of the *attracting* body. In this case, not only can the effect be calculated but the cause can be quantified in operational terms. This link between effect and cause, furnished by the two functions of mass, was, as we have seen, a major triumph of the theory of gravitation. Mass measures, not only the response of a body to an imposed force, but the force produced *by* the body on a second body. The other sorts of motion investigated by Newton—magnetic, chemical, vital—did not reveal any such link. And it would have seemed to him unlikely that matter of itself should turn out to have other active properties that would govern vital motions, for example, in the neat way that active gravitational mass governed the gravitational effects produced by bodies on other bodies at a distance from them.

The term 'govern' recalls our discussion of the hesitations he felt in attributing direct causal responsibility to the "attracting" body. But one thing he could be sure of: whatever the real *cause* of gravitational motion, the measure of the effectiveness of that cause was given by the mass-parameter of the attracting body. He could have no similar assurance about the other sorts of motion: though the forces involved in them could be computed from an observation of the motions themselves by means of the Second Law, no "laws of motion" proper could be determined in the absence of a quantitative measure of the *active* agencies involved. It was thinkable that such should be forthcoming for magnetic and electrical action:

> The attractive virtue of the whole magnet is composed of the attractive virtues of all its particles, and the like is to be understood of the attractive virtues of electrical and gravitating bodies. And besides these, the particles of bodies may be endowed with others not yet known to us.[161]

Even in the case of magnetic attraction, of course, there was no reason to suppose that the magnetic "virtue" could be thought of as a property of the matter of the body.[162] Newton notes that "the magnetic attraction is not as the matter attracted"; some bodies are attracted much more easily than others.[163] (Nor is it, of course, as the "matter attracting"; the mass of the magnet will not of itself determine what force the magnet will exert.) If a body is of uniform composition and is uniformly magnetized, then the more of it there is, the greater the total magnetic force it would generate, of course. In this sense, there would be a rough correlation between total magnetic attraction and the size of the magnet, between the intensity of the attraction produced and the amount of magnetic material. But this would not provide a mass-related parameter to describe the "magnetic virtue" of a particular magnet.

Newton had reason, then, to distinguish between two sorts of active principles, one which was matter-related and whose measure he had been able to construct, and the other which might be size-related but was evidently not invariantly related to matter as such. Because of his uneasiness about accepting the active power of attraction as a property of matter, he would not in fact have considered such a distinction a safe one. As far as he was concerned, it was more prudent to assume that even in the case of gravity the "causes" of the motion were still unknown; it was enough that he had been able to determine "two or three general principles of motion from phenomena" from which the properties and actions of bodies could be derived. He was happy to "leave their causes to be found out".[164] At the time he wrote this, he may still have had the "sensorium" metaphor in mind. If God moves all things after the fashion in which the human mind moves the parts of the human body, the "causes" of motion would seem to be permanently out of the reach of the inductive method, which could testify only to "effects". Were this to be the case, the provisional "positivism" of the *Principia* in regard to causes would turn out to be the best that the "experimental philosophy" could do.

§4.3   Light as a Mover of Matter

There was one sort of "active principle" to which special significance had been given by most of Newton's predecessors, especially those influenced by neo-Platonic doctrines. Light seemed to be the main source of life and energy for all types of organism; lacking the properties of solidity and inertia that characterize matter, it appeared to be intrinsically active. In his optical writings of the 1670s, Newton defended the view that light consists of a stream of corpuscles of some sort. But *what* sort? What properties do they posses? This could best be answered by studying the various ways in which light could be "inflected", or bent, from its original path in such phenomena as reflection, refraction, or dispersion. Descartes had already derived the sine law of refraction from the assumption that light travels at different velocities in different media; light could be regarded as a stream of material particles, following the same laws as any other material bodies.

But how is the change in velocity brought about? Newton first proposed a density gradient in the aether as the cause of inflection (1675), but was unable to find a consistent way to develop the optical consequences of this model. In the *Principia*, his new way of representing force led him to draw an analogy between a light-particle in a uniform field of force perpendicular to the interface between two media and a projectile moving under a uniform gravitational force perpendicular to the earth's surface. From this he was able to derive Snell's Law. The inflection would not be a sharp directional change, then (as it had been for Descartes), but a gradual parabolic curve. Not so gradual, though, by comparison with the paths of ordinary projectiles. Light can be bent through a large angle in a fraction of an inch. The forces involved, Newton reasoned, must therefore be enormous. In 1706, when preparing the Queries of the *Optice*, this clue led him to ask: "Since light is the most active of all bodies known to us and enters the composition of all natural bodies, why may it not be the chief principle of activity in them?"[165] And so he opened Query 22:

> Do not bodies and light mutually change into one another? And may not bodies receive their active powers from the particles of light which enter into their composition?

There are several suggestions here: one that light and material bodies are basically different in character, another that they can change into one another, and a further one that the activity of bodies derives ultimately from that of light. He

modified the latter in the 1717 *Opticks:* "may not bodies re-
ceive much of their activity from the particles of light which
enter into their composition?" leaving the way open to attributing at least *some* of the activity of a body to other causes. The
only points he actually *argues* for are (1) light enters into the
constitution of material bodies (they absorb and can emit light)
and (2) "the changing of bodies into light and light into bodies
is very conformable to the course of Nature, which seems
delighted with such transformations".[166]

He takes it for granted, both that light and body are somehow different in their basic nature (though he will also speak of
the rays of light as "very small bodies") and that light is more
likely to be a source of activity than body is. In computing the
"attractive power of a ray of light in proportion to its body",
Newton assumed that it is proportional to the velocity of the
ray and inversely proportional to the radius of curvature of the
deflection caused.[167]

> And by this proportion I reckon the attractive force of the rays of
> light above ten hundred thousand thousand millions of times
> greater than the force whereby bodies gravitate on the surface of
> this earth in proportion to the matter in them. . . . And so great a
> force in the rays cannot but have a very great effect upon the
> particles of matter with which they are compounded, for causing
> them to attract one another. Let us therefore see if they have not
> such attractions.[168]

He then goes on to the detailed account of the diversity of
ways in which the "small particles of bodies . . . act at a distance upon one another for producing a great part of the
phenomena of Nature", which makes up Query 31(23).

The mode of operation of this force is far from clear. Why
does it manifest itself suddenly at the interface? Does it operate
only between light-particles, or between them and matter-
particles too? If the latter, in virtue of what property of matter?
To say that light-particles exert a special attraction on matter

does not of itself imply that the measure of the responsiveness of the matter to this attraction would be the *mass* of the matter (as it would be were the attraction to be gravitational). Nothing in the *Principia* suggests what the correlative of gravitational mass would be in such cases. But what the *Principia* does specify is the inertial behavior of the light-ray according to the Third Law. If the ray exerts its force on something, then, by reaction, its own course is altered, and the ratio of force to mass-of-ray measures the *acceleration* of the ray. This is all that Newton needs for the qualitative picture he is trying to construct. The force differs from the gravitational force with which he is here contrasting it. And it continues to be exercised when the light-particles are absorbed; Newton surmises that it may be responsible, in part at least, for the short-range attractions by means of which the minute constituents of solid bodies hold together and interact with one another in such a diversity of ways.

He does not suggest that the presence of light-particles in "gross bodies" could also be responsible for the gravitational attraction exerted by these bodies. Yet to the extent that he entertained the possibility that light is *the* source of active power in bodies, he would presumably have had to ask himself whether the active force of light-particles could possibly explain gravitational attraction. One immediate counter argument would have been the strict proportionality of active gravitational mass and force. The gravitational power of a body to attract another depends only on its quantity of matter, and not (so far as one could see) upon the presumably contingent distribution within it of light-particles.

But now another question forced itself on Newton. Why *should* the light-particles exert the enormous forces they do? Several answers were possible. If the "stuff" of which the particles were composed were intrinsically highly active in a non-mass-related way (as Newton had noted magnets to be), there could be "light-forces", presumably obeying a distance law. Such forces, while differing from the simple mass-determined gravitational attractions of the *Principia*, could still be understood in terms of the general dynamical framework of that work.[169] Their inertial effects, the accelerations they produced, would be calculated in the same way as any other. In particular, the inertial mass of the light-particle would determine its reaction behavior under the action of the light-forces it exerted, even though its gravitational mass had nothing to do with the genesis of those forces.

Two other ways of generating the requisite high-intensity forces also occurred to Newton, and he mentions them more than once in his drafts for the later Queries. One is forces varying with distance by an inverse power greater than the square. The second is forces varying with the sizes of the particles involved. In *Principia* I, he noted that magnetic attraction is short-range and

> surely decreases in a ratio of the distance greater than the duplicate, because the force is by far stronger on contact than is the case with attracting bodies which are very little separated from their neighbours.[170]

In *Principia* II, he made this more precise, suggesting that magnetic attraction decreases approximately "in the triplicate proportion of the distance" (inverse cube). Around 1692, in some notes for the *Opticks*, he speculated that the cohesion of solid bodies may well be due to "attractive virtues" that decrease with distance by an inverse power greater than the fourth. These would decrease so rapidly that "great bodies composed of such particles shall not attract one another sensibly".[171] All of these suggestions illustrate the general theorem enunciated in *Principia* I, that forces which decrease very rapidly with distance were to be explained by laws of force involving powers higher than the inverse square.[172]

The rapid increase in intensity of such forces close to the attracting body does not derive from a special force–mass relationship; it depends entirely on the distance factor.[173] In the *Principia*, Newton showed that if a mass-proportional force varies with distance or with the inverse square of the distance, the total effect of the various parts of the attracting body is equivalent to that of the sum of their masses acting at the center of gravity in accord with the same distance-law. He regards this as a "remarkable" result, one which will not hold for higher inverse-power laws. However, it would be "tedious" he remarks, to run through them. The only theorems he derives for these other laws show, first, that "the attraction is infinitely increased when the attracted corpuscle comes to touch" a sphere which is attracting by an inverse third, or higher, power of the distance[174] and, second, that the distance law for a large body in such a case will not be the same as for its parts taken separately.[175] He does not discuss whether the additivity of mass also breaks down for such a body.[176] In summary, then, the enormous short-range forces apparently exhibited by light-particles might be explained by conventional force-laws, using powers higher than the square, without need for any other special postulate.

But this was *not*, in fact, the model that Newton most favored. The problem that had most preoccupied him in his earlier work in optics was chromatic dispersion. Why does white light, refracted through a prism, give rise to a spectrum? Newton argued that the different colors were already present in the apparently white light; their separation in the prism would thus be due to the differences in their refrangibility. But why should the corpuscles corresponding to the different colors behave differently at interfaces? If the corpuscular model were the appropriate one (as Newton was committed to showing), only a few parameters were available to explain these differences of behavior: velocity, size, shape, mass, elasticity, . . . Newton mentions the first three and adds conveniently vague terms, such as 'strength' and 'vigor', to denote the special force that light-particles might plausibly be supposed to possess.

The model he uses in the *Principia* to explain Snell's Law favored velocity as the differential factor. Mass would make no difference to the angle by which the light-particle was refracted. But the incident velocity of the particle *would* affect this angle. Velocity differences might thus (as Descartes had earlier suggested) suffice to explain dispersion phenomena. Newton, in fact, had used a velocity model in some of his earliest writings on color, but by the mid-1670s appears to have laid it aside.[177] He considered it more likely (by analogy with sound) that all the vibrations he was associating with the light-particles would travel with the same velocity, no matter what their "bigness".[178] The term 'bigness' and 'size' are, in fact, the ones that most often recur in this context in his notes at this time.

It was not, however, until he was drafting the Queries for the *Optice* in 1706 that he was forced to become more specific. In a draft for Query 21 he proposes:

> Colours and refractions depend not on new modifications of light but on the original and unchangeable properties of its rays, and such properties are best conserved in bodies projected. Pressions [pressures] and motions are apt to receive new modifications in passing through several mediums but the properties of bodies projected will scarce be altered thereby. Nothing more is requisite for the diversity of colours and degrees of refrangibility than that the rays be bodies of different sizes.[179]

The direction of his argument is clear. Whatever is responsible for color ought to be an intrinsic and permanent property of the light-particles, and not (like pressure of momentum) something liable to alter during inflection. And size, he says confidently, is all that the explanation requires. But how is this to work? A new idea finds expression in these drafts, and finally in Query 22 of the *Optice*:

> For the attraction of bodies of the same kind and virtue is the greater the smaller they are, in proportion to their size. This force is much stronger in smaller magnets than in larger, in proportion

to their weight. For the particles of small magnets, since they are closer to one another, unite their forces more easily. [180]

Although this passage is dropped from the *Opticks* of 1717, the inverse-size idea still carried weight with him at that time:

> Gravity in the surfaces of small globes is greater in proportion to the globes than in the surfaces of great globes of equal density. And therefore since the rays of light are the smallest bodies known to us... we may expect to find their attraction very strong. [181]

What are we to make of this argument? It rests on an apparent equivocation. The gravitational force exerted by a planet on a test body close to its surface is roughly proportional to the radius of the planet, and thus will *not* be greater for a smaller planet than for a larger one. What *is* greater for the smaller planet is the ratio of the force close to the surface to the volume (or to the mass) of the planet that produces the force. Newton's law gives us $F/M = m/r^2$; so the F/M ratio will increase as r decreases. If r is of the order of the radius of the planet, d is the density of the planet, and V is its volume, then $F/V = md/r^2$. So Newton is correct in saying that the F/V ratio increases for smaller values of r, that is, for smaller planets.

What is *wrong* with this is the implication that smaller planets are somehow more active, that they "produce" more force per unit volume than larger ones. The metaphor is admittedly seductive, but it is fallacious. To know how "active" a planet or a light-particle is, we have to measure the total force it exerts; its force/volume ratio does not yield an index of its activity, if "activity" is a matter of its effectiveness in moving a mass close to its surface. What the force/volume ratio *does* measure is the acceleration produced on the planet by the test-particle. So that, naturally, the smaller the planet (or the light-particle), the more it will be affected by reaction to a given test-body. But this in no way allows us to conclude that "since the rays of light are the smallest bodies known to us, we may expect to find their attraction very strong".

Nor does it allow us to conclude that such rays are more

affected in a given field of force than larger bodies would be. Newton toyed with an idea like this. If the rays of light are bodies of different sizes, he says in Query 29, the smallest would be the "weakest color", violet, which would be more easily diverted by refracting surfaces than would the larger particles, corresponding to the "stronger" colors up to red.[182] The suggestion is not that the smaller is the more active but that it is the more easily diverted. If the inflection is due to a uniform field of force (as in the *Principia*), then each particle would be accelerated to precisely the same extent. If, however, the inflection occurs because of the near approach of the light-particle to a matter-particle, then it would be correct to say that a smaller particle would be more "diverted" by such an approach than a larger one would be, simply because it can *get* "closer" (i.e., closer to its center of mass).[183]

This explanation does not attribute any special *activity* to light-particles. It derives from a "closeness" or "size" factor, and it is perfectly in accord with the laws of the *Principia*. But it is not of much service for refraction, where uniform forces perpendicular to the interface had proved their worth and where individual interactions between the light-particles and the particles of the medium would be far too complicated to compute. Nor is it of much service for cohesion; Newton's suggestion that the "small parts" will unite more strongly confuses attraction with the attraction/mass ratio.

The model Newton is using here can be called as "inverse-size" one. It ought *not* be called an "inverse-mass" model.[184] Mass is, of course, proportional to size if density is constant—so that Newton's argument *could* be stated, it would seem, in terms of inverse mass. If it holds for F/V, it will also hold for F/M. But one must note that Newton never uses the term 'mass', or an equivalent, in discussing this property of the light-particles. He speaks of size, of bulk, of magnitude; he lists all sorts of properties that might affect the attractive power of the particles, but does *not* list quantity of matter as one of them. This can hardly be accidental. The factor is clearly *size*-related, as such. The effect occurs not because the smaller particle has less mass but because it can get closer to the test-body.

It would have seemed paradoxical to Newton to argue that the less mass a particle has, the greater its attraction would be. Mass in the *Principia* is a measure of the active power of a body to attract. It is as though the matter is causally endowed with the power to attract, and the more matter there is, the greater the attraction will be. He could use an inverse-*size* argument, without this paradoxical feature's becoming too evident. The danger of representing his thinking by the term 'inverse mass' is that it suggests that in his later years he abandoned the force/mass proportionality that had been a cornerstone of the *Principia*, and which has remained a cornerstone of mechanics to this day.[185] There is not, however, the least trace of any weakening in his commitment to this principle. Because of the extensive evidence he had earlier brought in its support in the *Principia*, it would have been extraordinary had he sacrificed it for a speculative model that never *did* explain what it had been intended to explain. But more important: the inverse-size model (as we have seen) did not, in fact, depart from the principles of his earlier mechanics. A tension exists only if one insists on speaking of inverse *mass* (which he did not), and if one supposes him to refer to attraction rather than attraction/bulk ratio—a confusion for which he is himself in part to blame. It seems preferable, however, to attribute a confusion to Newton about the "virtue" of a particle (taking it sometimes to be the attraction exerted by the particle and sometimes the attraction/volume ratio), rather than to suppose that he implicitly, and for no good reason, abandoned one of the central principles of his entire mechanics.

Let us return, finally, to the question from which this section began: Can light serve to explain the motion of matter? It would seem that Newton's attempts to support an affirmative answer failed, but his work in optics must have strengthened his conviction that distance-related forces are the basic mode of physical explanation. It may also have helped make him more comfortable with the metaphor of forces operating across space, for inflection appeared to require them just as surely as gravitation did.

§4.4   The "Electric Spirit"

Around 1707, Newton became interested in electrostatic phenomena, as these had been demonstrated before the Royal Society by Francis Hauksbee. To his mind, they showed that bodies contain within themselves an "electric spirit" which could bring about the emission of light, besides causing small objects to move. Here, then, was a promising link between the phenomena of light and those of motion. The ontology he proposed was of a "certain most subtle spirit" (as he called it in the General Scholium of 1713) which pervaded solid bodies and the space immediately surrounding them.[186] Its effect would be to set up strong forces of attraction and repulsion between the smallest particles of bodies, thus giving rise to coherence, chemical change, growth, and generation and even, perhaps, "uniting the thinking soul and unthinking body".[187]

The advantages of invoking such a "spirit" (reminiscent, inevitably, of what would later be called electrical field) were manifold. It suggested how a wide range of phenomena could, in principle, be united by the same laws of force, analogous with the long-range gravitational forces of the *Principia*. And by assigning a *source* for the activity, it allowed a much more satisfactory explanation than gravitation had yet received; the only source Newton had been able to find in the latter case (as we have seen) was the direct action of God. The electric spirit implicitly pushed God's action a stage further back in chemical and other changes attributable to short-range forces, thus moving closer to the self-sufficient "natural" universe insisted upon by Leibniz. It permitted him to speak of forces exerted by bodies upon one another, without, however, making the constituent particles of matter a source of motion in their own right:

> Thus by the agitations of this spirit, to the extent that their constituent particles slip past one another more or less easily, bodies become soft and ductile or melt and change into liquid form . . . By the action of this same spirit, some constituent particles of bodies are enabled to attract one another more strongly, others less strongly, and thence can arise various gatherings and separations in fermentation and digestion. [188]

One is reminded of the ingenious hypothetical models in Descartes' *Meteorology*. But there is one essential difference: Newton has a theory of force which, in principle, would permit him to reconstruct these forces from postulated motions of the particles. The obvious explanatory value of the approach attracted him, it is clear. It had some inductive support already in the Hauksbee experiments, and it avoided the difficulties inherent in aether solutions, where aether is conceived as a medium for the transmission of action.

In light of this it is strange that, after numerous and quite detailed drafts of proposed new Queries concerning the electric spirit, [189] he should have decided to add nothing more on this topic to the cryptic paragraph in the General Scholium (1713), which ends on this cautionary note:

> These are things that cannot be explained in a few words, nor are we furnished with that sufficiency of experiments which is required to [sic] an accurate determination of the laws by which this electric and elastic spirit operates. [190]

Granted that this kind of restraint is to be expected in the *Principia*, why did he not develop the "spirit" model further in the Queries of 1717? He had found a way of representing the

action of all short-range forces, allowing him to range across
the entire spectrum of chemical and physiological changes—
over everything, in fact, except gravitation, the long-range
force, which would not easily lend itself to this sort of interpretation, unless one were to postulate a "spirit" that pervaded all
of space. What makes it all the odder is that he *did* introduce a
new model for gravitational and optical phenomena in these
Queries, an aether in fact, to which he even attributed the
"animal motions" earlier assigned to the electric spirit.[191]
Nothing more was said of this, but of course there is no reason
to suppose that he did not implicitly retain this sort of model
for short-range chemical phenomena to complement the explanation of long-range and optical phenomena in terms of an
aether.

§4.5 The Active Aether

This late return—Newton was now in his seventies—to an
aether theory at first sight appears inconsistent. After all, he
had long ago shown that an aether, involving contact action
through a *plenum*, would be necessarily "in exile from the
nature of things"[192] because of the resistance it would oppose
to planetary motion. How could he now propose a medium
that would by "impulse" explain gravitational action? If it
could exert an impulse, would it not also offer resistance to
motion? In a draft-letter reacting to Leibniz' views (c. 1714) he
suggested one way out of the dilemma: "a substance in which
bodies move and float without resistance, and which has therefore no *vis inertiae* but acts by other laws than those that are
mechanical".[193] And he went on to defend the propriety of
non-mechanical modes of explanation in general, for, after all
even "Mr. Leibniz himself will scarce say that thinking is
mechanical". But will this work? Had he not earlier asserted,

in the strongest possible terms, that a non-material medium, altogether lacking in gravity, would not be a "phenomenon" and thus would not fall withing the scope of experimental science?[194] It would be of little avail to point to Hauksbee's electrostatic experiments.[195] The electric spirit could come much closer to qualifying as a phenomenon than would a medium lacking in *vis inertiae* and thus not governed by the notions of force defined in the *Principia*.

This may have been why the aether Newton *did* propose in 1717 was of very small, but still finite, density. It was postulated, in the first instance, to explain the vibratory phenomena of light and the phenomena of radiant heat (demonstrated by the transmission of heat through a vacuum). Newton initially had enough confidence in this proposal to think of including it in a series of "Observations" (rather than putting it in query form).[196] He was much more tentative, however, about his attribution of a gravitational role to the new aether. Presumably, there was no question of returning to a Cartesian *plenum*, which in a note from this period he still castigates as having "no place in experimental philosophy".[197] In Query 22, he maintains that a fluid with no "pores" would be denser than quicksilver; vacuities were essential, in his view, to explain density differences, and since the density of the interplanetary medium would have to be extremely low, it would have a large proportion of void.

It does not operate, then by contact action. Two other possibilities appear to have suggested themselves. One was a mode of action akin to that of the electric spirit.

> There are, therefore, agents in Nature able to make the particles of bodies attract one another very strongly to stick together strongly by those attractions. One of those agents may be the aether above-mentioned, whereby light is refracted. Another may be the agent or spirit which causes electrical attraction. . . . And as there are still other mediums which may cause attractions such as are

the magnetic effluvia, it is the business of experimental philosophy to find out all these mediums with their properties.[198]

Here the aether would simply endow material particles with the properties of attraction or repulsion, properties that matter could never (in Newton's view) of itself possess. It would not, therefore, be a "medium" in the ordinary sense, nor would it be material or a body (as he defines these terms). One might be able to attribute "density" to it in some metaphorical sense, alluding (presumably) to the intensity of the forces it sets in train, but there would by no reason to think of it as particulate.[199]

In the published text, however, Newton tries a rather different model: an aether consisting of particles that repel one another strongly, thus giving the medium an enormous "elastic force" allied with very low density. He takes the speed of transmission of vibrations to depend on the square root of this ratio,[200] and estimates that the known velocity of light would require an elasticity/density ratio at least 49.10^{10} greater than that of air. The density of the medium is supposed to increase with distance from the nearest bodies; "if the elastic force of this medium be exceeding great, it may suffice to impel bodies from the denser parts of the medium towards the rarer, with all that power we call gravity".[201] The explanation of how these density gradients operate is treated as a further and more speculative suggestion. Newton inserts an extra qualifier, "if anyone should suppose", before introducing the hypothesis that the aether consists of particles that repel one another, and adds "for I do not know what this aether is". He is much more confident in propounding a vibratory aether than in making it particulate-repulsive in order to function in gravitational explanation.

One might well ask what advantage Newton thought to gain by an aether which still required forces operating at a distance, in the same way as the long-range gravitational forces he is trying to eliminate by their means. A possible answer suggests itself: that he could allow *intrinsic* activity to the aether more easily than he could to ordinary matter. To accept such an aether would avoid the troublesome implication of his gravitational theory that matter is essentially active, the doctrine we have seen him fight so hard against. In draft-material of this period, Newton tries a distinction:

> To distinguish this medium from the bodies which float in it and from their effluvia and emanations and from the air, I will henceforward call it aether, and by the word bodies I will understand the bodies which float in it, taking this name not in the sense of the modern metaphysicians but in the sense of the common people, and leaving it to the metaphysicians to dispute whether the aether and bodies can be changed into one another.[202]

Elsewhere in this same material he says that anything is body which can be transmuted into what is commonly called "body". That he should leave it to the "metaphysician" to decide whether this is true of aether indicates that he does not think that there is an empirical way of settling the issue—or perhaps that he wants to avoid it.

His dilemma can be simply stated: if the aether is material, with a finite *vis inertiae* (as the density and vibration metaphors, as well as the velocity calculations, suggest), then to attribute inherent force to its particles would be to concede to at least *some* matter (and perhaps, therefore, to all matter, if his transmutation views hold good) an intrinsic activity. Matter itself would become the source of new motion in the world. On the other hand, if the aether is non-material, lacking in *vis inertiae*, how can it be a "phenomenon"? And if it is *not* a phenomenon, it can "have no place in experimental philosophy"—a principle he emphasized elsewhere in this same draft-material.

His uncertainty about where to position aether in the spectrum between mechanical medium and active immaterial spirit is illustrated in this draft passage from the same period:

> Body I call everything tangible which is resisted by tangible things. . . . Vapors and exhalations on account of their rarity lose almost all perceptible resistance, and in the common acceptance often lose even the name of bodies, and are called spirits. And yet they can be called bodies if they are the effluvia of bodies and have a resistance proportional to density. But if the effluvia of bodies were to change thus in respect of their forms so that they were to lose all power of resisting and cease to be numbered among the phenomena, these I would no longer call bodies, for I speak with the common people.[203]

Here the matter–spirit boundary seems to depend not on the intrinsic dynamism that is characteristic of spirit but upon the "power of resisting" that is characteristic of matter. Something that lacks this power, that is, *vis inertiae*, is no longer a "phenomenon".

Let us recall that he has been trying to counter those who (like Leibniz) criticized the hypothetical elements in his system, by retorting that he admits no such elements and that his arguments are entirely inductive. The debate with Leibniz ought to have made him realize that his criterion for an entity's qualifying as a "phenomenon" (and thus as a proper subject of experimental science) was too narrow. If phenomena must exhibit *vis inertiae*, then non-inertial active principles must be relegated not just to the "queries" of a natural philosopher but to "metaphysics and hypothetical philosophy", domains in which he professed no interest whatever. It is curious that someone who invested so much creative effort in these spirits and aethers—to the point, indeed, of making them, with their attendant forces, the explanatory basis of his system (subsequent to 1707, at least)—should at the same time, under pressure, have allowed the concession that only entities which resist impressed motion are legitimate topics for the natural philosopher. It is fortunate that his practice belied his theory in this regard, else eighteenth-century science would have been deprived of some of its most significant clues.[204]

§4.6 Rational Reconstruction

How are we, finally, to map the structures of inference and interpretation in Newton's attempts, subsequent to 1687, to provide explanatory models of the active principles of motion? What led him to seek and pursue these alternative models with such tenacity? Let us try the following reconstruction, keeping in mind how fluid and tentative and shifting the process was, like any other prolonged creative process.

(1) Newton accepted the necessity of going beyond the mathematical formalism of the *Principia* to give a causal account of gravitational and other motions. A straightforward positivist ascesis might suffice as a short-term device, but it was certainly inadequate as a definitive position for someone who regarded himself (as he certainly did) as a natural philosopher.

(2) Two quite different guiding principles suggest themselves as ways of understanding the line his speculations took:

(P1) Bodies cannot act at a distance upon one another without the aid of an intermediary.[205]

(P2) Matter cannot of itself be the source of new motion.[206]

(3) It follows on the basis of either P1 or P2 that apparent motions at a distance can be explained only by postulating the agency of non-material entities (e.g., direct divine action, non-material entities operating at a distance, a non-material medium) to transmit action.

The first point to note is that P2 is broader than P1; in fact, P1 can be inferred from it. If matter cannot initiate activity, then (accelerated) bodily action at a distance (or for that matter contact action giving rise to accelerated motion) is impossible.

So if Newton's fundamental premise was the "inertia" of matter, the action-at-a-distance prohibition would have followed. In the letter to Bentley, however, Newton seems to suggest P1 as the more fundamental:

> It is inconceivable that inanimate brute matter should, without the mediation of something else which is not material, operate upon and affect other matter without mutual contact, as it must if gravitation . . . be essential and inherent in it.[207]

Thus he moves from the unacceptability of bodily action at a distance to the unacceptability of a gravitationally active matter. Yet in later draft-material, and even in published texts, Newton often appears to take the action-at-a-distance model quite seriously. Query 31, for instance, opens:

> Have not the small particles of bodies certain powers, virtues, or forces by which they act a distance, not only upon the rays of light for reflecting, refracting and inflecting them, but also upon one another for producing a great part of the phenomena of nature?[208]

Indeed, the entire direction of his thought is to explanatory models that *do* invoke action at a distance (as do all four of his suggestions: divine action, active principles as forces, electric spirit, active aether). Of course, they do not call for bodies *as such* to act at a distance; this is precisely the too-simple solution that they are intended to neutralize. But where is the negative emphasis here? Is it on bodies *acting* (initiating new motion) or on their acting *at a distance?* Would the immaterial agencies have been needed to explain a contact action that generated new motion? It is interesting to note that Newton assumed *all* new motion to come about by the operation of agencies at a distance; contact can only redirect motion, the sum of the momenta being conserved. It is not clear why he could not have had active principles at work in the *impact* of bodies, as both Descartes and Leibniz had suggested. Perhaps the mechanical metaphor was too strong, or perhaps he could see no reason to suppose that new motion ever *did* come about in this way. It seems clear, in any event, that the "at-a-distance" aspect was not the crucial one.

Space, in short, was no longer the inefficacious vacuum of the mechanical philosophers; rather, it was filled with vaguely stated entities of all sorts. In the 1690s, he had explicitly made the divine action fill all space. Later, when he turned to the electric spirit, the spaces immediately surrounding electrified bodies were assumed to be pervaded by it; whereas the optical aether filled all of space, though leaving "pores". Space thus appears as the arena of spirit, as that in which spirit characteristically acts to move matter. To the extent that forces are somehow associated with the operations of spirit, they too could be thought of as in some vague sense existing "in" the intervening spaces. However, he never clearly situates them as entities in their own right, bridging the gaps between gravitating bodies.[209] Yet there can be little doubt that the metaphors of spirit and active principle, with their overtone of a mode of presence in space other than that of material occupation, helped to dissolve the doubts he had expressed about action at a distance.[210]

Can we take it, then, that the dominant consideration in Newton's mind in his elaboration of the various explanatory alternatives was P2, the principle of the passivity of matter? This seems much closer to the mark, for the reasons we have seen above. Newton believed that it had inductive warrant and, clearly, it continued to influence him, as he sought one means after another to avoid attributing active agency to matter. Nevertheless, it was, as we have seen, compromised on occasion not only by the uncertain immateriality of the obviously active aether particles and "effluvia" of various sorts but even by his appearing to attribute forces to body directly (as in the opening sentence of Query 31 above), without any explicit mention of a separate active agency responsible for them.[211] So P2 cannot be construed as a guideline that he single-mindedly pursued.

Another possible clue to a consistency of argument might be the theological rationale that was partly responsible (as we have seen) for Newton's insistence on the passivity of matter, namely, the undesirability (in his eyes) of allowing matter to appear as altogether self-sufficient in its activity, because of the resulting implicit exclusion of God from His universe. Yet this interpretation will not do either; the various sorts of active principle he invoked more and more from 1706 onwards, though perhaps not material, were certainly not identical with God's direct power. To see *them* as the sources of change in the world made the world no less self-sufficient than directly attributing the origin of activity to matter would have done. Of course, the active principles were "spirit", in some broad sense, but *not* in the sense that their agency was reducible to that of God.

We have failed to identify any single principle, consistently pursued, as the driving force behind Newton's search for an explanatory account of gravitational motion. Can we turn, perhaps, in a methodological direction and suggest that he is simply seeking models of high explanatory value?[212] The electric spirit and the active aether, after all, are constructs with *some* potential for generalization and unification of data.[213] On reflection, however, this seems inadequate (on its own, certainly) to account for the course things took. First, such notions as spirit and active principle of themselves gave neither a mathematical nor a physical "hold" on the phenomena. What mattered were the *forces*, and it did not make much difference, from the standpoint of inductive confirmation, how the ontology of these forces was sorted out. From a broader explanatory standpoint, of course, it *did* make a difference, but here we are back where we started: when we ask *why* they had explanatory value, we have to invoke such principles as that of the passivity of matter. The optical (not the gravitational) aether had a rather more autonomous status; as we have seen, the model functioned as more than a simple carrier of forces in this instance. But this is in a different context from mutual action at a distance.

Another difficulty for the "methodological" account is suggested by Newton's return to an aether solution in 1717. Why did he not, in the end, simply concede that all bodies could be endowed with powers of gravitational attraction, both active and passive, as the text of *Principia* seemed to say quite unequivocally? Surely, this would have been more consonant with his methodological principles than putting up for consideration a theory with such implausible features (e.g., an aether whose density *increased* with distance from matter and whose "elastic force" was fifty million times greater than that of air)? What could have motivated him to sail so close to the wind?

The only plausible answer would seem to be that acceptance of mutual attraction would contravene both P1 and P2. True, the successive editions of the *Principia* spoke freely of such attraction, but Newton clearly hoped to find a more fundamental ontology underlying it, one which would make the attractions only *seeming* attractions, in the sense that the ultimate source of gravitational motion could be attributed elsewhere and the laws of "attractive" force could be retained unchanged. He had a precedent; his early attribution of short-range forces to the primordial particles of matter had later found satisfactory amplification in explanatory models (the electric spirit, active principles, the aether), which shifted active agency from matter itself. Could the same be done for the long-range forces of gravitation? His first attempt, as we saw, was to call on the metaphor of soul and body to suggest that bodies are moved by God directly, according to laws that operate as though force were being attributed directly to the bodies. His second attempt was the aether of 1717. Neither idea was especially successful.

Their importance for us lies not in their success (or lack of it), however, so much as in the clue they offer to the implicit criteria that govern Newton's quest for explanation in mechanics.

Back in 1693 he had declared to Bentley that no man who has "in philosophical matters a competent faculty of thinking" can ever fall into the "absurdity" of claiming that brute matter can act at a distance.[214] It would be too much to claim complete consistency for him in regard to this bold declaration (so paradoxical-sounding on the lips of the man who had just based a celestial mechanics on "attraction"), but it seems plausible to argue that the conviction of an "absurdity" lingered with Newton and that it impelled him to reach beyond the descriptive statements of the *Principia* to an explanatory model which did not confer activity on matter, particularly not the ability to act gravitationally upon other matter at a distance.

§4.7 Conclusion

After inquiring into the universal properties Newton attributed to all matter, we fastened on three interconnected questions that continued to plague him from the publication of the *Principia* onward: Is matter active? Is gravity an essential property of matter? How is matter moved? His answers to all three of these were governed, in part at least, by concepts of matter inherited from a multiplicity of philosophical traditions. But for none of the three was a straightforward answer possible; there was a tension of a new kind between the demands of a successful, mathematicized physics and the intuitive intelligibilities of common sense. It is instructive—and fascinating—to follow the zigzag paths of Newton's speculative genius as he sought, without much real guidance from the past, to resolve this new sort of tension.

In retrospect, it may be helpful to distinguish three different sorts of interaction between Newton's concept of matter and his mechanics. There was, first, the notion of a "quantity of matter", measured by the product of density and volume, which finds echoes in the nominalist physics of fourteenth-century Paris, and which Newton transformed for his own purposes to unify the contexts of resistance to change of motion and of gravitational attraction. The amount of "stuff" in the body became a triple measure of inertia, of attractive power, and of power to be attracted; it was defined, not just by Definition I, but by the eight Definitions and three Laws, taken as a single conceptual network.

The way Newton chose to "mesh" quantity of matter with the other mechanical concepts was confirmed not so much by the new celestial mechanics (where it played no direct part) as by the successful treatment of falling bodies and, specifically, pendulum motion.[215] The sharpening of the intuitive notion of a quantity of stuff came in consequence not of philosophical reflection as such, nor of new observations, but of an effort to bring the observational data already available under a single system of dynamical concepts. All of these were drawn from ordinary usage but were articulated much more precisely in order to satisfy the descriptive, predictive, and explanatory demands of a mathematicized and experimentally based mechanics. There was a definite continuity between the *quantitas materiae* of the older natural philosophy and the newer concept; however, the development from one to the other was governed at least as much by the technical constraints imposed by previously known empirical laws in astronomy and mechanics as by the fruitful intuitive metaphor of a featureless "stuff", the metaphor that had contributed so much to the earlier tradition.

A second and quite different influence of Newton's concept of matter on his mechanics, can as we have seen, be traced back to neo-Platonism, alchemy, and the hermetic assumptions he found so congenial. Its effect was to "liberalize" New-ton's ontology and to encourage him to move beyond the rigid reductionism of the mechanical philosophy. Thus he could feel at home with the concept of attraction in his mechanics and introduce the metaphors of power, spirit, and principle into the natural philosophy that was prompted by this mechanics. Even more important, he was led to look for non-conservative sources of motion in the world; neither the *materia* of the Aristotelian or nominalist traditions nor the matter of the mechanical tradition would have suggested to him that he should do so (except in the domains of life or mind).

Thus the new mechanics evidently drew part of its inspiration from this complex and heterodox source. Certain sorts of things made sense that would not have made sense for the mechanist; explanatory entities were introduced whose value would become apparent only in the later history of chemistry and electrical theory. Though Newton could not escape from the reductionism implicit in the primary–secondary distinction, his thought was free where agency and hypothetical entity were concerned. His creative genius showed itself in the wide-ranging expansion of the ontology of material agency almost as much as in the development and application of powerful new mathematical techniques.

Finally, there is the complicated story of material action at a distance. Newton's strong belief in the passivity of matter was rooted in the older natural philosophies, particularly neo-Platonism. The new mechanics seemed to be leading him, almost inexorably, to a modification of this principle. Yet he could never quite bring himself to accept this. Though his entire gravitational theory hinged on the notion of mutual attraction, he tended either to take this term in a purely descriptive sense or to try to find explanatory agencies underlying attraction, agencies that would not inhere in matter as such. The intensity of his quest for these agencies, manifested publicly in the Queries and attested to privately in the copious draft-material from his later years, cannot but impress us. It is, without question, one of the most instructive case histories on record of the interaction between the metaphysical and the physical dimensions of the great scientist's thought.

CHAPTER FIVE

Epilogue: Matter and Activity in Later Natural Philosophy

NEWTON'S LEGACY TO his followers in regard to the relations of matter and activity was hardly a clear doctrine. Rather, it was a set of suggestions, of metaphors, of rather tentative principles. No wonder, then, that the Newtonian "school" diverged in so many directions in regard to this issue. There were three major challenges to the notion of matter that it had inherited and there was no unanimity as to how they should be met.

In their discussions of primary and secondary qualities, both Boyle and Locke had emphasized that matter must be conceded to have certain "dispositions" or "powers" to act upon mind, else it would not be perceptible.[216] These powers derived, in their view, from the primary qualities; nonetheless, an objective status had to be acknowledged for them. It is a fact that gold produces a visual sensation of a certain sort in man. The power to do this is dispositional (it is not always in actual exercise), and it is relational, in the sense that the presence of a certain sort of organism is required for its exercise. But it is nevertheless intrinsic to gold as a power, and just as objective a fact as is its possession of mass.

Thus Newton's insistence on the inertness of matter in its own right raised problems of an epistemological sort. Would such matter be perceptible? And if it were not, and only the forces or agents ar active principles that surround it were capa-

ble of acting, why postulate matter in the first place? Here the
ambiguity in Newton's manner of associating matter and force
(or power) became critical: either the matter was somehow
responsible for the force, or they were merely associated in
some external way. If the former, then the matter itself would
have to be allowed to be the source of activity, to possess
powers in its own right. If the latter, it was unclear why the
forces should depend on the parameters of the matter in the
first place—why, for example, the greater "force" Newton associated (as we have seen) with smaller particles should in fact
depend on the size of the particle. The occasionalism implicit
in Newton's natural theology came close to the surface here:
did God, in His wisdom, simply correlate forces of a certain
strength with pieces of inert matter of a certain size in a purely
external way? If He did, how could one *know* that He did?

As early as 1712 a Cambridge natural philosopher, Robert
Greene, began to formulate this difficulty in an obscure and
cumbrous work, *The Principles of Natural Philosophy*, which
was explicitly designed to repudiate some of the central ideas of
that earlier Cambridge work of almost the same title.[217] He
rejects, on empiricist grounds, the doctrines of a void space
and homogeneous matter of uniform density. He insists that
matter must be thought of as active, otherwise it is no more
than an abstraction. The main characteristic of matter is *resistance*, and resistance is due to the exertion of a resisting force
on the part of matter. This leads him to say unequivocally that
action or force is the "essence of matter"[218]—about as direct a
repudiation of the Newtonian principle as one could imagine.

His theory of matter proceeds, he tells us, "upon this one
plain axiom, that it is impossible for us to have any sensations
from matter but by some kind of action or other impressed
upon our minds from it."[219] Matter "without one real force
belonging to it", possessing only the primary qualities of the
corpuscular tradition, is no more than a "fanciful and humor-
ous abstraction" for whose existence we have no real war-

rant.[220] Matter, endowed with "expansive" and "contractive" forces, suffices to explain not only the phenomena of gravitation, magnetism, and so forth but also the so-called "primary" properties of extension and solidity.

Thus Greene is attacking on two fronts at once. From the epistemological side, he asserts that matter must at least be allowed the power to affect sense (a theme afterward developed more skillfully by Reid); from the side of physics, he urges that a great simplification can be worked by regarding *all* properties of matter as the effects of forces intrinsic to it (a theme that will find its fullest expression later in Boscovich).

A second challenge came from an idea that Newton himself had been the first to formulate.[221] If light is to find "easy passage" through transparent substances, these substances must contain only a very small proportion of solid parts (which "stifle" the rays that fall on them).[222] Thus matter, despite its appearance of solidity, must be porous. The tenuity of matter was strongly stressed in a passage added to the 1717 *Opticks*. Newton's disciple, John Keill, had already inferred that the quantity of matter in a piece of glass might have no greater proportion to the bulk of the glass than a grain of sand to the bulk of the earth.[223] More significantly, and more perilously, Clarke had applauded Newton's mathematical philosophy as the only way to prove that matter is the "smallest and most inconsiderable part of the universe", a doctrine that gave Leibniz grave offense.[224] This was to its theological credit, Clarke assumed, because it enlarged the domain of spirit, which in turn exhibits God's action in a manifest way.

But if the less solid matter there is, the more God's power is manifest, an obvious question arises: Is there a lower limit to the tenuity of matter in sensible bodies? Newtonians, such as Pemberton, Desaguliers, and Martin, kept pushing the tenuity thesis till Voltaire, in 1733, could remark that we are not certain "there is a cubic inch of solid matter in the universe, so far are we from conceiving what matter is".[225] This qualifica-

tion would not have pleased Newton. who was fairly sure he
knew what matter is, but it underlines the increasingly problematic role of a vanishingly tenuous matter that is quite unlike (it would seem) the hard, massy, extended atoms of the
corpuscular philosophy.

The most radical challenge, of course, came from Berkeley. His target was not just P2, the principle of the inertness
of matter; he rejected matter of *any* sort, active or passive. He
was especially concerned to exclude the forces and agencies
that populated the Newtonian universe, whether or not they
inhered in matter. He did not share Newton's confidence that
such forces are a means of entry for God's agency. Berkeley saw
more clearly than Newton that the matter–spirit dichotomy
ceased to be a means of ensuring the universality of God's
action in the world once the domain of "spirit" was governed
by the quantitative laws of physics. In the older, narrower
sense of 'mechanics', Newton might argue that his active principles were "non-mechanical". But Berkeley saw (as later
eighteenth-century Newtonians were to see) that the *Principia*
had given grounds for extending the sense of the term
'mechanical' to apply to the apparently "immaterial" agencies
of gravitation and electricity. In Berkeley's theological perspective, to call these "immaterial" in no way safeguarded the
primacy of the divine action; they were ultimately the powers
of material body in Newton's system, part of the physical order
and not properly to be identified with the direct operation of
God. Berkeley thus brought out the ambiguous status of that
"spirit" realm which was, for Newton, at once both material
and immaterial, the arena of measurable physical forces, as
well as somehow giving assurance of the penetrability of the
natural order to God's immediate action.

He agreed with Newton that division of reality into inert and
active sectors was crucial to a sound natural theology. But
where Newton had made only matter inert, Berkeley extended
the claim of inertness to the entire phenomenal realm. Adopting Locke's ambiguous usage of the term 'idea', he rejected

Locke's claim that idea-things have the power to affect the mind:

> All our ideas, sensations, or the things which we perceive, by whatsoever names they may be distinguished, are visibly inactive; there is nothing of power or agency included in them. . . . Whoever shall attend to his ideas, whether of sense or reflection, will not perceive in them any power or activity; there is therefore no such thing contained in them. A little attention will discover to us that the very being of an idea implies passiveness and inertness.[226]

Idea-things cannot be the cause of sensations, then, nor can they affect one another. The notion of attraction or force "signifies the effect, not the manner or cause". Natural philosophers discover regularities of succession in the phenomena, as well as "analogies and harmonies between them"; but causes lie forever beyond their reach for these "can be no other than the will of a spirit".[227] The ordered succession of ideas is the "inert matter" of Berkeley's system, but Berkeley can deny the label 'matter' to it (and thus eliminate this term from natural philosophy entirely) by asserting the dependence of "idea-things" on mind. If that dependence be denied, of course, Berkeley's system, in the eyes of its Newtonian critics, will become an invitation to atheism: the dividing line between matter and spirit is gone, and there is no longer any warrant from within physics for the existence of a "spiritual" order.

Berkeley's attack on the assumptions concerning cause and explanation that underlay Newton's work, and indeed all seventeenth-century natural philosophy, was later carried much further by Hume. It had relatively little effect on natural philosophy itself, where talk of forces, powers, and agencies continued unabated. However, it must have served to reinforce doubts about the theological warrant that had influenced Newton and so many others in adopting a matter–spirit dichotomy as basic to natural philosophy.

Newton's attempt to maintain this dichotomy and the resulting principle of the inertness of matter was thus undercut from several quite different directions. Nevertheless, it retained an appeal for many. John Hutchinson strongly defended it in the 1720s on theological grounds. Bryan Robinson, writing in 1743, constructed an ingenious aether theory, utilizing hints from Newton's later work. Aether is construed as the cause of gravitational motion, as well as optical and electrical phenomena. Matter could thus be taken to be inert, and "the cause, which gives the aether its activity and power, must be spirit".[228] The "imponderable fluid" theories which were gaining adherents at this time tended to locate the source of agency primarily in the fluid (caloric, phlogiston, electricity).[229]

What contributed, perhaps, more than any direct challenge to the decline of P2 was the very language of Newtonian mechanics. As terms like 'attract' and 'force' became more and more familiar usage in mechanics, Newton's strictures against taking them in their customary sense tended to be forgotten. The artificiality of the separation between matter and force needed a strong motivation for its maintenance, and as the earlier motivation declined, natural philosophers took these terms at their face value and spoke of bodies "acting on" one another, without any of the reservations that Newton had tried to enforce. It was much more economical to allow that bodies *could* act; one could still, if one wished, invoke a variety of "agencies of Nature" in the space surrounding matter, but one did not have the puzzle of correlating these agencies with a matter on whose properties they apparently depended but which was supposed to be incapable of affecting them causally.

Boscovich's influential *Theory of Natural Philosophy* (1763) is particularly interesting in this regard. The author tells us that he is trying to bridge the gap between Newton and Leibniz by postulating point-centers of force instead of extended solid corpuscles. The points possess inertia, and can properly be called "matter". Extension, solidity, and gravitation are to be explained by a single law of force which is alternately attractive and repulsive, depending on the range. The matter-points "float" in a vacuum. Electricity involves a fluid (he accepts Franklin's theory), and magnetism *may* involve "some intermediate kind of exhalation, which owing to its extreme tenuity has hitherto escaped the notice of observers, and such as by means of intermediate forces of its own connects also remote masses."[230] But, in general, he does not fill space with forces or powers; the problem of action at a distance does not appear to worry him.

His notion of force is hedged with the same reservations Newton had expressed earlier. Force is no more than a propensity of matter-points to approach, or recede from one another:"The term does not denote the mode of action but the propensity itself, whatever its origin, of which the magnitude changes as the distances change".[231] Despite his use of such terms as 'attraction' and 'act on', Boscovich is trying to avoid attributing agency directly to the matter-points.

> Whether this law of forces is an intrinsic property of indivisible points; whether it is something substantial or accidental superadded to them . . . ; whether it is an arbitrary law of the Author of Nature, who directs these motions by a law made according to His Will; this I do not seek to find, nor indeed can it be found from the phenomena, which are the same in all these theories. The third is that of occasional causes, suited to the taste of the followers of Descartes; the second will serve the Peripatetics. . . . The first theory seems to be that of most of the modern philosophers, who seem to admit impenetrability and active forces, such as the followers of Leibniz and Newton all admit, as the primary properties of matter founded on its very essence. This theory of mine can indeed be . . . adapted to the mode of thought peculiar to any one of these kinds of philosophizing.[232]

This striking passage is almost Berkeleyan in discounting any warrant from phenomena for a theory of agency. His unwillingness to commit himself, as a natural philosopher, to an account of the *causes* of matter's propensity to move in lawlike ways is not inspired by Newton's motives. Instead of taking the agencies responsible for gravitational and other motions, accounted for by the notion of force, to be somehow the testimony of the presence of "spirit", and thus indirectly (perhaps) of God, Boscovich is expressing a skepticism based on epistemological considerations in regard to all such moves.

His own matter-spirit dualism (on which he sets great store) rests on a quite different basis and has a quite different function. It is aimed at safeguarding the position of *man* in a mechanical universe, rather than at pointing to God directly. Spirit is that which is responsible for vital activity of all sorts. It is not restricted to mind, as Descartes held, but extends to *all* living action. Spirit, since it can properly be said to move matter, exhibits agency in the fullest sense. The "forces" in corporeal matter are "nothing else but propensities to local motions . . . and these do not depend on any free determination of the matter itself", unlike the operations of spirit.[233]

> There will not be any great difficulty over the use of terms, so long as matter (which is devoid of all power of feeling, thinking and willing) and living things possessed of feeling are carefully distinguished from one another.[234]

Though Priestley was much influenced by Boscovich, it was this matter–spirit distinction that he set out to combat above all else in his *Disquisitions Relating to Matter and Spirit* (1717). "Whatever matter be, I think I have sufficiently proved that the human mind is nothing more than a modification of it".[235] Since the solidity of matter can be explained by repulsive forces, there is no need to invoke the solid corpuscles of traditional Newtonian thought. Nor ought one distinguish between the force and the matter of itself "destitute of powers" lying behind it; the epistemological argument against this is decisive. The tenuity argument ought to have warned natural philosophers long before that if solidity had so little to do in the system, "perhaps there might be nothing for it to do at all".[236]

Instead of making powers "accompany and surround the solid parts of matter", why not *equate* matter with powers or forces (he does not distinguish the two)? Or, more exactly, why not define matter as "substance possessed of the property of extension and of powers of attraction and repulsion"?[237] If one does this, there is no longer any reason to think of sensation and thought as outside the possibilities of matter, just because they are active agencies. They are simply further powers among the many, some known and others unknown, that constitute matter in its essential nature.

Priestley thus goes much further than Boscovich. He does not just endow matter with forces, he *equates* "matter" with forces, thus rejecting the older notion of matter entirely. His is thus a one-level universe, just as Berkeley's had been, but it is the category of matter, rather than spirit, that fits it best (though it has to be modified to do so). Gone is the matter–spirit dichotomy of Newton; gone even the less problematic dualism of Boscovich. Most radical of all, the distinction between matter and force, which is at the root of Newton's entire theory of natural agency, is, if not eliminated, of only minor significance at best. Force inheres in matter; it is essential to it. Without it, matter vanishes. It is still described in terms of property. But it is constitutive of the being of matter; there is no ground for distinguishing a matter-substance from its force-activity.

With Priestley, the pendulum has swung a very long way from Newton. There will still be those like George Adams, who in his *Elements of Natural Philosophy* (1794) harkened back to the solid inert corpuscles of the older tradition, but most will follow the example of Priestley, at least to the extent of attributing agency directly and unreservedly to matter.[238] In his *Metaphysical Foundations of Natural Science* (1786), Kant gave the fullest philosophical expression to this view, arguing in great detail that matter of its nature exerts both repulsive and attractive force. The former allows it to occupy space, that is, gives it extension; the latter is necessary for the existence of matter as an organized complex. And attractive force is defined as that "whereby a matter can be the cause of the approach of other matter to itself".[239]

Kant contrasts the "mechanical way", which combines "the absolutely full with the absolutely empty", with the "dynamical way", which explains "all varieties of matter through the mere variety in the combination of the original forces of repulsion and attraction".[240] He repudiates the former, the "mechanical philosophy" of extended corpuscles moving in an empty space, which rests, he believes, on the assumption that density differences can only be explained by supposing internal vacuities in matter. This assumption leads, he notes, to the familiar tenuity thesis that the "filled" part of even the densest matter might be "very nearly nothing compared with the empty part"; Kant evidently regards this as implausible. The "dynamical philosophy", on the other hand, can explain these differences by repulsive forces of different degrees, and this explanation is to be preferred as simpler and more fruitful because it leads to a larger doctrine of forces.[241]

Kant proved to be wrong in this estimate; Dalton was soon to show the utility of extended atoms. He could (and did) claim a Newtonian warrant for them. But on a broader front, Newton's program for chemistry, to which so much effort had been devoted in the decades after his death, did not prove successful. Instead of quantifying short-range forces, the new chemistry quantified atomic weights, of all things; the key would be laws of combinations, not laws of attraction. The ambitious Boscovich/Priestley proposal proved not to have the explanatory resources that chemistry and physics needed at that juncture; they had moved too quickly to generalize from Newtonian mechanics to natural science as a whole.

In all of this, one can see how the general categorial scheme a scientist adopts guides his research. When one says that the principle of the inertness of matter was ultimately abandoned, it might sound as though this were an empirical discovery of a straightforward kind: scientists working with matter simply found it to be active. But, of course, it was not quite like this. Rather, what was at issue was how the concept of matter should be used and, specifically, how it should relate to the other general cosmological categories of spirit, power, force, space. Before commenting further on this, perhaps we ought to turn briefly to the other principle whose influence on Newton we traced above: the inadmissibility of action at a distance (P1). Here Newton's legacy proved rather more beneficial to his heirs, though perhaps even less definite than in the case of P2.

We have seen that four different strategies were possible in regard to the notion of attraction. One could simply propose that A acts on B at a distance. One could speak in terms of propensities and say that A has a propensity to move in a certain way, given the presence of B, prescinding, for the moment at least, from the agencies responsible for the motion.

One option, then, was to accept the notion of action at a hold that such agencies are in principle outside the reach of the natural philosopher, for whom, then, the question of action at a distance would be vacuous. Or one could rely on hypothetical media, fluids and the like, as a means of transmission of action from A to B. Newton felt the first of these strategies was unacceptable, and would not have thought the third any more congenial. In practice, he relied most on the second, but obviously felt the fourth to be the most desirable, when it is available.

All four strategies are to be found in the writings of eighteenth-century natural philosophers, but two developments ought especially to be noted. The success of the Newtonian account of gravitation appeared to give a *prima facie* warrant for the possibility, if not the plausibility, of action at a distance. The conceptual incoherence that many of Newton's generation found in the notion disturbed scientists less and less as they became accustomed to the new model. Its predictive power served to persuade many that P1 ought not, at the very least, be regarded as an absolute principle of natural philosophy. Kant went further and argued the proposition that "the attraction essential to all matter is an immediate action through empty space of one matter upon another".[242] His grounds are that since contact is due to attractive force, contact action cannot be made a condition of operation for such force. Thus its operation is "independent of the filling of space" between the two bodies. The argument, that matter cannot act where it is not, he undercuts by noting that no body can act strictly where it *is:* in all material action, the agent acts on something outside itself, that is, where it is not.

He continues:

> It is commonly held that Newton did not find it necessary for his system to assume an immediate attraction of matters, but with the strictest abstinence of pure mathematics herein left the physicists full freedom to explain the possibility of such attraction as they might find good, without mixing up his propositions with their play of hypotheses. But how could he establish the proposition that the universal attraction of bodies ... is proportional to their quantity of matter, if he did not assume that all matter exercises this motive force simply as matter and by its essential nature?[243]

One can find two strands of argument in this and the following passages. The first is that "true attraction" *would* produce an effect on B proportional to the quantity of matter in B; since this is what is observed, it must be due to true attraction. (This is not a necessary consequence, as he assumes it to be). The second is that the alternative of a fluid medium operating through contact action begs the question, since such media also operate by means of forces of repulsion and attraction. He concludes that "the offence which his contemporaries and perhaps he himself took at the concept of an original attraction made him at variance with himself".[244] Only a true attraction, not an apparent one explicated in terms of impact and intervening media, can explain the phenomenon of approach. Kant believes that Newton was asserting true attraction. Newton was right to leave aside the *cause* of this attraction (Kant argues), for these fundamental forces are assumed as basic and cannot be further derived:

> The concept of matter is reduced to nothing but moving forces. . . . But who claims to comprehend the possibility of fundamental forces [i.e., explains them causally in terms of a yet more basic principle]? They can only be assumed, if they inevitably belong to a concept which can be proved to be a fundamental concept not further derivable from any other. [245]

One option, then, was to accept the notion of action at a distance as basically intelligible and, in fact, as the *preferred* mode of explanation in physics. But a second and, in the long run, more influential tendency was to construe the forces themselves as extended entities, filling the space around bodies. There was, as we have seen, more than a suggestion of this in Newton's own thinking about active principles. It was not difficult to move from consideration of an "aether" or an "atmosphere", conveying the effects of force, to a reification of force itself, of the kind we have seen in Priestley and others. Euler's work on the mechanics of fluid media was of fundamental importance in this respect, even though Euler himself opposed such reification. He provided an idealized mathematical model for the transmission of action in a continuous medium, thus enabling those who thought of force as something propagated across space to interpret this image in a precise theoretical way. In fact, Euler's work formed the basis of a new metaphor, that of a *field*, understood roughly as a region of space in which each point is characterized by quantities which are functions of the space and time co-ordinates.

But is there any real difference between such a field and straightforward action at a distance, in the absence of a fluid for which there is independent evidence? This was a question that greatly exercised early nineteenth-century physics, but it would take us too far afield to follow it in any detail.[246] Faraday was the first to try to meet it head-on by formulating criteria for the interpretation of a particular field as involving real processes in the intervening space. If the velocity of propagation is finite, or if it is affected by what goes on in the intervening space, or if it does not depend on the parameters of the "receiving end", there is reason to think that real intermediate agency is involved. On the basis of these criteria, optical, electrical, and magnetic fields can be taken to be "real", whereas the gravitational field qualifies as action at a distance. Maxwell suggested a further indicator: the detectable presence of field-connected energies in the intervening space. And he unified light, magnetism, and electricity by means of a single formalism that appeared to qualify for an "aether" (or at least *some* sort of real agency) as its referent. Yet all attempts to find an aether of a mechanical sort failed; action-at-a-distance theories (which had remained popular in France and Germany) gained support from this apparent failure of the British continuous-action approach. And positivists, such as Hertz, could reproach both sides for their foolishness in clothing the mathematical formalism with the "gay garment" of a physical interpretation.

The advent of relativity theory complicated the issues even further. The finite velocity of propagation, imposed on all forms of energy transmission, counts in favor of continuous-action models. And the gravitational field theory involves energy flow and a "medium", space-time itself, whose structures are determined by the presence of matter. But the original clear choice between continuous action and action at a distance is now obscured. The "intervening empty space" no longer appears in the picture and the interpretation of the "action", described by the energy-momentum tensor in realist terms, presents difficulties. Nonetheless, relativity theory is generally held to favor continuous-action over action-at-a-distance principles.

Our purpose in this brief review was the limited one of indicating how complex the subsequent history of principles P1 and P2 has been. In the case of P1, the end of the story has assuredly not yet been written. There can be no denying the importance of the part these principles and others like them played in the thought of Newton and those who followed him. Should we allow the positivist claim that science would have been better off without this excess metaphysics, and that the history of science has witnessed its gradual and continuing elimination? Can we concede Hertz' distinction between a neutral mathematical formalism, which is the "real" science, and its theoretical interpretation in a physical model serving only as an aid to imagination?

The story of Newton ought to be enough, of itself, to show the inadequacy of these no longer so fashionable views. It is easy to "freeze" theory at an instant, focus on its "hard" component (usually either in mathematical or operational terms), and treat the remainder as a dispensable crutch for the imagination. But this omits the role played by such physical notions as attraction, or such principles as P2, in the construction of the theory from which the formalism is a later, and artificial, abstraction. It is all very well to say "Let us not ask what 'attraction' means beyond the operational facts of mutual approach." But if Newton had not had the physical notion of attraction to begin with, and had not considered it a plausible way of describing the planet–sun relationship, the mathematical formalism of gravitation would never have been set down. It is the *theory* that grows and develops, the *theory* that counts as explanation. Even where the formalism can be given three different interpretations (recalling Boscovich), the real question is: Which of these "interpretations" (the word itself is a

bad choice since it suggests something subsequent) enabled the
formalism to be arrived at in the first place? And which now
gives it the most resilience in the face of anomaly? the most
resources for further development? and so forth.[247] In practice, it is the theory as a whole which is assessed by the scientist,
not the abstract skeleton of mathematical equations or operational formulae.

Newton tried in vain to restrict readers of the *Principia* to
formal considerations. But the shape of what was to come
cannot be found in that formalism. We had to look for the
principles underlying his groping attempts to construct a co-
herent *explanatory* account of motion. It made all the difference in the world to his heirs whether or not action at a
distance, for example, was to be regarded as an acceptable
explanation. The entire course of their theorizing, as we have
seen, was quite likely to be affected by this option. It was not a
speculative interpretation added after the event to a formalism
already experimentally established, then, but a principle guiding the construction of theory and its later modification in the
face of challenge.

Newton's own metaphysics was not particularly successful,
as we have seen. P2 met sharp criticism from epistemological,
methodological, and metaphysical quarters, and had to be
abandoned. P1 was attacked on metaphysical grounds and has
survived only in drastically modified form. But we must be
careful not to confuse one question, Did the principle of the
inertness of matter further Newton in his development of a
good theory of motion? with another question, Did reliance on
metaphysical principles further Newton in his work generally?
A negative answer to the first question would in no way commit one to a negative answer to the second. Newton could not
have developed his theories without metaphysical principles of
some sort.

There had to be decisions about where to make the cuts in
Nature, about where to seek causal agency, about what should

count as explanation. Such decisions went a long way beyond the inductive warrant afforded by earlier successful science. There was more open space in Newton's day; there were more decisions to be made on a prescientific basis. We may be tempted to deplore, for example, his constant recourse to a natural theology which too readily drew God into physical agency. But he had to turn to what he regarded as his surest insights, and this would have counted among them.

The experience of several centuries has served to eliminate principles that once influenced the course of science and to give others the sanction of success. One might be tempted to think that regulative principles of a broadly metaphysical kind no longer play a role in the natural sciences. Yet even a moment of reflection about the current debates in elementary-particle theory, in quantum-field theory, in cosmology, ought to warn that this is far from the case. True, the principles at issue may not be as overtly metaphysical as they often were in Newton's time; but the distinction is one of degree, not kind.

The complexity we have found in Newton's lifelong attempts to formulate the basic explanatory concepts of his science is perhaps the best indication of the creative genius that was his. The task was much harder for him. That it should be easier for us, we owe in part to the wealth of unruly suggestions he left behind.

Notes

[FULLER BIBLIOGRAPHICAL DETAILS and a key to the abbreviations will be found in the References, below.]

Introduction

1. *Principia*, Book I, Def. VII. Page citations of the *Principia* will be to the Motte-Cajori translation. See Mandelbaum, *Philosophy, Science and Sense-Perception*; pp. 66–88. The intellectual antecedents of many of the themes to be discussed below are traced in "The concept of matter in transition," my introduction to *The Concept of Matter in Modern Philosophy* (Notre Dame, Ind., 1978).

2. *Principia*, Book I, Prop. 69, Scholium.

3. Newton hesitated between two quite different notions of explanation, one deducibility from a covering law and the other the discovery of causes. A study of the *Principia* and the *Opticks* reveals two different clusters of terms, one having to do with derivation and demonstration and the other referring to causes and reasons. The tension between these two quite different ideals of science pervades Newton's work. See Anita Pampusch, C.S.J., *Isaac Newton's Notion of Scientific Explanation*.

4. Book III of the *Opticks* remained unfinished. Newton claimed that his work had been "interrupted" but it is more probable that he could not generate an adequately based theoretical account of the diffraction experiments he describes. The sixteen queries appended to the first edition dealt with the interactions of matter and light (diffraction, light emission, generation of heat, vision, color dispersion). Seven queries (17–23) were added to the *Optice* of 1706; their theme: "are not all hypotheses erroneous in which light is supposed to consist in pression or motion, propagated through a fluid medium?" (Q 20). The final query (23) of this edition was longer than all the

others put together; it is our main source for Newton's chemical theories, specifically for the "powers, virtues or forces" by which the "small particles of bodies" act upon one another to produce not only optical phenomena but all the vast array of chemical changes. Newton added eight new queries to the 1717 edition; they reintroduced the old notion of an aether, modified to meet some of the objections Newton himself had raised in the earlier queries and in the *Principia*. These were numbered 17–24, and the queries which had been added to the second edition of 1706 (17–23) were renumbered 25–31. We shall write these as, for example, Q 31(23), indicating the 1706 numbering in parentheses. See Bechler, *NLF*, pp. 187–8.

5. *Opticks*, p. 339. It is, in fact, in the queries (Q 28[20]), in the midst of a particularly speculative passage, that Newton unexpectedly sets down the oft-quoted maxim that "the main business of natural philosophy is to argue from phenomena without feigning hypotheses and to deduce causes from effects till we come to the very first Cause, which certainly is not mechanical".

6. The various editions of the *Principia* will be designated by Roman numerals. Thus "*Principia* II" is the second edition.

7. It is not altogether clear to what extent (if any) Clarke can be taken to be a spokesman for Newton in this debate, as D. T. Whiteside points out in a personal communication. There is no direct evidence of their having discussed the issues. Of course, they were both residents in London at this period, so that meetings between them would not necessarily have left written traces. Yet it is notable that there is no evidence in Newton's voluminous correspondence of any briefing. When Leibniz wrote to the Princess of Wales in 1715, criticizing Newton's views, she communicated the letter to Clarke directly; he had translated the *Opticks* into Latin and was known to be concerned with the implications of Newton's mechanics for natural theology, the main topic of Leibniz' original letter. A. Koyré and I. B. Cohen have argued from four indirect manuscript clues that Newton was involved behind the scenes in the controversy, and their argument has been widely accepted as conclusive ("Newton and the Leibniz–Clarke correspondence," *Arch. Intern. Hist. Sciences*, 15 [1962], 63–126). It is, however, unclear whether Newton took an active part in the drafting (or correcting) of Clarke's replies, as against merely knowing the contents of the letters after they appeared. Koyré and Cohen themselves conclude: "We have found no evidence to make precise the degree of either Newton's participation in the Leibniz–Clarke correspondence or of Clarke's participation in Newton's letters to Conti" (p. 79). In what follows, then, we will not assume that Clarke's contribution to the *Correspondence* can be interpreted as Newton's own doctrine conveyed by an official spokesman.

8. See Cohen, *PL*, and Hall and Hall, *USP*. I owe a special debt of gratitude to J. E. McGuire, who has transcribed (and where necessary translated) many of the crucial passages dealing with issues in natural philosophy in a series of recent acticles (see References).

Chapter 1:   The Universal Qualities of Matter

9. *Principia*, Book I, Prop. 69, Scholium.

10. Book III, Prop. 6, Hypothesis 3.

11. See McGuire, *TI*, pp. 84–89, and Kubrin.

12. Second paper on light and colors of 1675, *PL*, p. 180.

13. In the *Principia* (Book III, Prop. 6, Cor. 4), for example, he infers the necessity of granting the existence of a void between the constituent particles of solid bodies from the postulate that these particles all have the same density. Jammer (pp. 67–68) argues that Newton cannot have held this postulate, but his supporting reference to Cajori (p. 638) is to a passage where Cajori is arguing that Newton did not assume the constituent particles to be of the same *size*, and his reference to the *Opticks* (pp. 403–4) is to a passage where Newton is asserting God's freedom to "make worlds of several sorts in several parts of the universe" with *different* "laws of nature" involving, among other possibilities, constituent particles "of different densities". There is, of course, the familiar charge of circularity against Newton's manner of defining mass in terms of density, even though density measurement appears to presuppose mass measurement. In the circumstances, it is not surprising that he had little to say about the density of the primordial particles, but he was clearly aware of the possibility of explaining *macroscopic* differences of density and kind without postulating similar primitive differences at the level of the primordial particles.

14. Although he never explicitly excepts the primordial particles from his assertions about universal transmutability, there is persuasive indirect evidence that he regarded them as unchanging. See McGuire, *TI*, pp. 81–84.

15. *USP*, p. 341. There is a similar passage in a draft preface for the same edition (*op. cit.*, p. 307), suggesting the importance he attached to this Hypothesis when composing the original text of the *Principia*. Cohen infers from the fact that Newton later described Hypothesis III as having been "directed against the prejudices of the Aristotelians and Cartesians" (*HNP*, p. 176) that he had never really believed it himself and that its function from the beginning was *ad hominem*. But in view of the other references above and the role played by transmutability in his thinking generally at that time, it could also be argued that he had advanced the view in *Principia* on his own account, had almost immediately realized the problem it entailed, and therefore proceeded to characterize it as "Cartesian" for the purposes of the argument in Corollary II, now transformed into an *ad hominem* argument only. See McGuire, *TI*, for fuller references to Newton's views on transmutability. See also footnote 122 below.

16. *Opticks*, p. 400.

17. ULC, Add. 4005, f. 81v (translated). For the manuscript evidence on the gradual evolution of the Rules of Reasoning, see Cohen, *FN*, and especially *HNP*, pp. 168–79; McGuire, *TI*, pp. 71–84.

18. Book III, Prop. 6, Cor. 2.

19. ULC, Add. 4005, f.81r.

20. The Rules were inserted at the beginning of Book III. The Latin phrase *"quae intendi et remitti nequeunt"* is a technical one, going back to the late medieval theories of quality change. It has a more active connotation ("which cannot be intended or remitted") than the customary English translation conveys.

21. ULC, Add. 4005, f.81r.

22. See, for example, McGuire, *EQ*; Koyré, "Gravity an essential property of matter," *NS*, pp. 149–63. In the earlier draft already quoted, Newton infers from the fact that they are not subject to alteration of intensity that impenetrability, inertia, and mobility can be taken "to belong to the essence of matter". But the terms 'essence' and 'essential' are dropped from the final version of the Rule (except where gravity is said not to be an "essential" property), and only *universality* is claimed for the properties earlier described as "essential". Presumably, he had noticed that universality was, in fact, all that the Rule entitled him to claim.

23. Translation by Hall and Hall. *USP*, pp. 360–61.

24. *Principia*, p. 546. Is it significant that the earlier draft was *not* published as it was? Here, as in many other instances in this paper, due weight must be given to nonpublication. When Newton altered or dropped a passage, one must obviously be cautious about attributing to him the views it contained. The context requires careful scrutiny in each case.

25. Rule III, p. 400; Introduction to *Principia*, p. xxvi.

26. Letter to Bentley, *PL*, p. 302.

27. 'Universal' is, of course, close to the vague sense in which the term 'essential' is frequently used today, but it could be misleading to describe Newton's doctrine in terms of 'essential" (rather than "universal") qualities since the term meant so much more for him.

28. The distinction is both epistemological and ontological in its thrust. See my introduction to *The Concept of Matter in Modern Philosophy*, where the different types of primary–secondary distinction are discussed.

29. McGuire notes a precedent of sorts in Digby's *Two Treatises*, a book Newton is known to have read, but argues that the emphasis and thrust given this criterion by Newton has no parallel in other writers of his day (*EQ*, pp. 244–46). Newton needed it to make a distinction between two types of phenomenal property, not between substance and quantity (*dimensio terminata*) on the one hand and quality on the other, which was the force it had had in the Scholastic tradition. The incoherence in its usage by Newton, noted below, was in part due to this transfer of context.

30. ULC, Add. 4005, f.81r. See McGuire, *EQ*, p. 257; Cohen, *HNP*.

31. M. Clagett, *The Science of Mechanics in the Middle Ages* (Madison, 1959), pp. 205–14.

32. See J. Bobik, "Dimensions in the individuation of bodily substances," *Philosophical Studies* (Ireland), 4 (1954), 60–79. In the long

drawn-out medieval discussion on the principle of individuation of material things, it was debated whether things were individuated simply by the possession of dimensions, that is, by dimension considered in abstraction from any particular "termination" or boundary (i.e., the property of extendedness) or by the possession of *particular* dimensions which would distinguish one body from another.

33. ULC, Add. 4005, f.81r.

34. *Principia.* p. 547. The phrase "rendering general by induction" occurs very frequently in his writings from this period. See Mandelbaum, p. 77.

35. The most helpful treatments of this topic are McGuire, AAN, and Chapter 2 of Mandelbaum, "Newton and Boyle and transdiction."

36. AAN, p. 3.

37. Mandelbaum (*loc. cit.*) uses "transdiction" for what we have called "transduction" above. In transduction, the only problem about "speaking across" is that we are going from the observable to the unobservable. To assume that a univocal language will suffice in such a case is to set aside the main obstacle that any attempt to "speak across" from the observable to the unobservable must face. We are using the term 'transdiction' here, not of all cases of "speaking across" but only of those where the difficulty of constructing an appropriately different mode of speech for the new and unobserved domain is specifically taken into account.

38. Edleston, *Correspondence*, p. 155.

39. Mandelbaum distinguishes between two levels of generality: "further cases of the same sort" and "applicable throughout nature" (pp. 77–78). I am not sure that there is a real distinction here, but in any event I am in agreement with his claim that Newton intends induction to warrant the latter as well as the former.

40. *Principia*, p. 400; for a similar maxim, see *Opticks*, Q 31(23), p. 404.

41. *Opticks*, pp. 403–4.

42. ULC, Add. 3965.6, f.256r. See McGuire, AAN, p. 29.

43. See §2.7 below.

44. There may be a parallel between Scotus' attempt to argue against Aquinas that at least one predicate (being) must apply univocally to all things, God included, and Newton's attempt to find univocal predicates to cover the entire realm of Nature and thus guarantee the "foundation of all philosophy".

45. Though Newton uses the phrase 'analogy of Nature' to describe the extension of the concept of vibration from acoustics to optics (McGuire, AAN, p. 37), he is not thereby committing himself to saying that there is no univocal concept that carries across. Though the term 'vibration' itself is analogously used of the two cases, the mathematical analysis that is characteristic of waves applies in the same way to both, and it is this univocal formal "core" that Newton is seeking to identify. To extend a concept "by

analogy" from a familiar context to an unfamiliar one does not as yet commit one to whether it should be applied univocally or only analogously (i.e., admitting elements of difference as well as "likeness") in the new context.

46. *Opticks*, Q 31(23), p. 376.

47. ULC, Add. 3970, f. 338v. These instances are taken from McGuire (AAN, pp. 38, 46), who in a detailed treatment reads them as being based on the conception of "a chain of bodies imperceptibly changing into one another" with properties in a continuum from one level to another (p. 40). The validity of the transdiction would depend, therefore, on the continuity of the gradations from one level to the next. I have argued, on the other hand, that 'consonance' conveys sameness, and that the thesis of gradual transmutation, far from being the basis of the "analogy of Nature", is precisely the doctrine which the "analogy of Nature" in Rule III is intended to limit more explicitly than before.

48. Draft of Rule III (ULC, Add. 4005, f. 81v). The wording here strongly suggests a regulative principle, a presupposition one is required to make in order for science to be possible, rather than a knowledge claim of a straightforward sort.

49. See Cohen, *HNP*, pp. 173–74.

50. *Principia*, p. 397.

51. Prop. 6, Cor. 3. Hypothesis/Rule III and Prop. 6, with its Corollaries, were evidently designed from the beginning with the Cartesians in mind, though the reference to them is sharpened, as we have seen, in the second edition.

52. *USP*, p. 316. This is from a draft for the projected edition of the *Principia* in the 1690s. The main consideration, of course, was density differences which, in Newton's view, were only explicable—since he evidently believed the primordial particles to be of uniform density—in terms of greater or lesser proportions of internal vacuity. In this draft, he estimated that the ratio of vacuum to solid matter in water—despite its incompressibility—is 65:1. See footnote 13 above; McGuire, *BV*; and Thackray, *MD*.

53. ULC, Add. 3970, f. 479r. See McGuire, AAN, p. 26.

54. In the early draft of Rule III (ULC, Add. 4005, f. 81r–v), he remarks that impenetrability is usually described as "belonging to the essence of bodies", and from this it is inferred to be a universal quality of body. But he objects to this manner of arriving at the conclusion. Since, according to him, "the essential properties of bodies are *not* revealed by the natural light" of reason, one must begin rather from the fact that impenetrability is universally possessed by the bodies we observe and proceed inductively from there. He is clearly rejecting the conceptualist/deductive method of Aristotle and Descartes here in favor of an empiricist/inductive one. The impenetrability issue had already been debated by Descartes and More. The former held that "impenetrability belongs to the essence of extension" (though body is still defined solely by means of extension, impenetrability being only a derived

feature), whereas More argued that impenetrability and extension were two separate features of bodily substance, with no necessary link between them. See Gabbey, p. 8.

55. One recalls in this connection the well-known passage in the *Saggiatore* where Galileo distinguishes between two sorts of property, one (such as extension, location, state of motion) which cannot be separated from the notion of matter "by any stretch of imagination", and the other (such as color and taste) which our minds do not recognize as necessary to materiality.

56. Later Newtonians will note that the introduction of force allows us to explain these properties as effects of force; we do not, therefore, have to postulate them as independent qualities. See Chapter 5 below.

57. Newton is not, therefore, defending an ontological distinction of the Galilean sort, which would make secondary qualities fictive or at best subjective. The title of Book II, Part III, of the *Opticks* is: "Of the permanent colors of natural bodies." He explains these colors by the ways in which the constituent particles of the bodies reflect and refract light, after the fashion of the thin transparent plates whose optical behavior he had so successfully explained. The color of a gold coin is thus just as relational, in his view, as that of a soap bubble, but it is a real disposition of the coin, to be explained by the shapes, sizes, and dispositions of the small constituent particles of the coin. The distinction between Newton's universal qualities and all other qualities is both an epistemological distinction of a weak sort (only the universal qualities are *necessarily* associated with matter) and a reduction distinction, since they are proposed as a basis for the explanation of all other properties.

58. So too did the fact that the universal qualities were not directly given in experience but were indirectly defined by their place in a network of explanatory concepts. Since this network is hypothetical and evolving, the "reductionism" it permits can never be of the definitive sort that the primary–secondary distinction of the seventeenth century and the various materialisms of the eighteenth century envisaged. But Newton could hardly have anticipated this consequence of his methodology.

59. *Opticks*, Q 31(23), pp. 401–2.

Chapter 2:   Is Matter Active?

60. See, for example, the draft-material for Query 31(23) (ULC, Add. 3970, f.619r). One is reminded of the central thesis of Aristotelian mechanics: "*quidquid movetur, ab alio movetur*" (whatever is in motion is moved by something other than itself), which implies an inability at all levels of nature to initiate motion and keep it going. In this view, only the Unmoved Mover is fully and truly active. See D. Shapere, "The philosophical

significance of Newton's science," in R. Palter, ed., *The Annus Mirabilis*, pp. 285–99; McGuire, *FAN*, pp. 164ff.

61. See Guerlac, *Newton et Epicure*; McGuire, *FAN*.

62. Ambiguous in its causes, and especially in its causal relations with matter and spirit. See Westfall, *FNP*, chapter 2; Koyré, *NS*, chapter 3.

63. See R. Blackwell's essay in McMullin, ed., *The Concept of Matter in Modern Philosophy*.

64. In "The principles of nature and grace," *Philosophical Papers and Letters*, ed. L. Loemker (Chicago, 1956), vol. 2, p. 1034.

65. See the useful discussion of these different types of force in Westfall, *FNP*, Chapter 6: "Leibnizian dynamics."

66. When using a mass factor, *m*, in speaking of Leibniz' mechanics, it is important to note that it is still no more than a quasi-intuitive measure of "quantity of matter", proportional to weight. A theory of gravitation is needed to convert it into a fully operational factor, and this Leibniz could not develop since he made impact the central paradigm of his mechanics, just as Descartes had done.

67. Postscript to the Fourth Letter to Clarke, in Alexander, p. 44.

68. 'Matter' and 'body' are usually interchangeable for Newton, but in this passage 'matter' is used when he is stressing inertia, 'body' when the emphasis is on impulse and the exercise of force.

69. 'Impulse' translates '*impetus*' here; since the impetus is said to be "exerted" and since the dynamic metaphors underlying the Definition are clearly drawn in the first instance from the domain of impact phenomena, Cajori's choice of 'impulse' seems apt.

70. The reader is referred to Cohen, *NSL*, for a full treatment of these passages and the difficulties in interpreting them. See also McGuire's Comment in the same volume, Ellis, and Westfall, *FNP*, Chapters 7 and 8.

71. It is fairly clear that Newton did not advert to these difficulties. He was using constant small increments of time which permitted him to ignore differences in order of infinitesimals; changes in numerical factors did not matter since he was dealing in proportionalities only. See Whiteside, *BNP*, and Cohen, *NSL*, for a discussion of this tangled topic.

72. Westfall remarks that familiarity has dulled our perception of the paradox in this strange conception that Newton bequeathed us: "If matter is endowed with an inherent force by which it resists efforts to change its state, it cannot be wholly inactive, or in the classic phrase of the seventeenth century, wholly indifferent to motion. Indeed, Newton's formulation of the laws of mechanics implied a conception of matter at once inert and active, unable to initiate any action itself, passively dominated by external forces, but endowed with a power to resist their actions" (*FNP*, p. 450).

73. It could be represented either as the source of the series of impulses or as the series itself, depending on ontological preference.

74. Their resemblance is due to the fact that both are held to affect motion by means of impulses represented as *mutationes*, instantaneous

changes. By making the impulses smaller and smaller, Newton assumes that he is approaching the case of continuous motion. Whether he was as aware of the mathematical complexities of this move as he is sometimes credited with being seems debatable. The basic problem here is that the VC impulses conserve motion whereas the VI impulses alter it; a parallelogram method cannot properly be used, therefore, to relate them. Furthermore, though impact, rather than continuous force, is the root metaphor of the mechanics of the *Principia*, it is not clear what the relation of *vis insita* is to impact changes. VC functions to conserve the motion of A, not to alter the motion of B, when A and B collide. If the *vis insita* of A is supposed to alter the momentum of B at impact, it has the same sort of duality that *vis inertiae* has (as we shall see below).

75. This is the sort of paradox that Westfall's interpretation of Newton seems to encounter: "From the point of view of the second law, an unbalanced force expends itself in generating a new motion. From the point of view of the third law, an unbalanced force is an impossibility. . . . In the one case, the equilibrium of force has as its result the maintenance of an inertial state. In the other case, changes in the inertial state re-establish equilibrium between the *vis inertiae* and the impressed force. From Newton's point of view, indeed, the perspective of the third law was the more fundamental. His dynamics concerned itself primarily with the relations of two forces, internal and external force, *vis insita* and *vis impressa*, and the third law stated their relation in terms of the ancient tradition of static equilibrium" (*FNP*, p. 470). This is to take Newton's notion of *vis* much too univocally, if we are correct. The notion of an "equilibrium" between *vis impressa* and *vis insita*, and the correlative distinction between "balanced" and "unbalanced" forces, can hardly help but suggest that *vis insita* is a force in the sense of Law II. If Law III were really the fundamental perspective that this "equilibrium" approach suggests, it is hard to understand how the entire first section of the *Principia*, dealing with central-force orbits, could have functioned without invoking it, as indeed it does. It is only in section 13 (p. 164) that the assumption of an "immovable center" of force is laid aside.

76. From this it cannot be inferred that the absolute quantity of the *vis insita* could be known, thus making it possible to distinguish between inertial frames. Newton believed that the analysis of rotational motion would allow him to discriminate between non-inertial frames, and thus in principle decide on the absolute quantity of *vires impressae*. But there was no similar device available for separating inertia frames; there was no way to know whether a *vis insita* was operating or not, since, as he pointedly notes in the Definition, "motion and rest, as commonly conceived, are only relatively distinguished". But the "conserving" language of Definition III, though it commits him to assigning a real conserving function to *vis insita*, no more commits him to the possibility of an absolute measurement of *vis insita* than the language of Law II *of itself* commits him to the possibility of an absolute measure of *vis impressa*. See McGuire, *CC*, p. 189.

77. See Cohen, *NSL*.

78. The basic duality of *vis viva* and *vis mortua* in some respects reflects the tension between Newton's *vis insita* and *vis impressa*. But *vis viva* was a more versatile and effective notion than *vis insita* proved to be. Leibniz took the ontological ramifications of mechanics more seriously than Newton did. His notion of a "formal effect" of motion perhaps came closest to the *vis insita*. See Westfall, *FNP*, p. 293.

79. See McMullin, ed., *The Concept of Matter in Greek and Medieval Philosophy*.

80. Though it might perhaps be identified with its effect in the other body (the impressed force). This would simplify the scheme! It depends on how the notion of *conatus* or effort is construed. If the *conatus* of B to change the motion of A is something predicable of B. then it is distinct from the force acting upon A. The notion of *conatus* in Leibniz is quite different: the *conatus* in B has to do with the motion of B itself, not of A.

81. McGuire maintains (as we do) that there are "two irreducibly different kinds of 'explainers' in the *Principia*: external impulses, which are causes for Newton, and an internal *vis insita* in each bit of matter". Inertial motion requires explanation, according to the *Principia*, and this is the role of the "internal, inherent and innate 'property' of all bodies" Newton calls "*vis insita*". McGuire argues that it is a disposition, absolute or "nongeometrical" in nature (independent of the positions and motions of other bodies). *Vis inertiae* (which McGuire calls the "state of *inertiae*") is an actualization or "manifestation of this hidden property", which is evoked by interaction with other bodies, and is thus "geometrical" (dependent on the existence and action of these other bodies). See his *CC*, pp. 189–190.

82. This is the main difficulty about taking VR as a sort of mirror image of VI, as for example Gabbey does. *Vis inertiae*, considered as resistance, is for him nothing more than VI from another point of view. Our reading of *vis inertiae*, understood *strictly* as resistance (i.e., as occurring in the body in which it can properly be called resistance), has led us to separate it rather sharply from VI. If A acts on B, the VR in B acts in opposition to the VI in B. The notion of a *vis impressa* does not of itself possess the duality which would allow it to be taken also as resistance, even in the context of mutual gravitational attraction, the context which is the source of most of the trouble.

83. Which Gabbey identifies as "the proper sense of *vis inertiae*" (p.47).

84. Westfall, *ANC*, p. 203. This article gives a good summary of Newton's alchemical work. A more detailed treatment may be found in B. J. T. Dobbs, *The Foundations of Newton's Alchemy*.

85. *Opticks*, Q 31(23), p. 397.

86. Q 31(23), p. 399. In the first version of this Query in 1706, he suggested the Divine Will as an alternative to "active principles" in this context, but eliminated this reference in the 1717 edition.

87. They are partly internal to *compound* bodies, since they operate be-

tween the primordial particles. Such bodies have various sorts of chemical and optical activities going on within them, but the active princples are regarded as extrinsic to the primordial particles themselves, and in that sense not inherent to matter as such.

88. Deleted passage in a draft dating from around 1705 for Query 23 of the *Optice*; ULC, Add. 3970, f.620v.

89. Though some would doubt their significance for the understanding of Newton's achievement: "Newton is alive not in the barren speculations of the Queries, but in the solid permanent mathematical pages of the *Principia*". From them, "our forebears learned how to use the concept of force given *a priori*" (C. Truesdell, "Reactions of late Baroque mechanics . . . ," in *Annus Mirabilis*, R. Palter, ed., pp. 226, 230). The reader will already have gathered that we reject this assessment; the grounds for our rejection will become more evident in the pages that follow.

On the other hand, we would not go as far as Westfall, who argues that if we take into account Newton's preoccupation with alchemical writings, hundreds of which he worked through carefully in the decade before the writing of the *Principia*, we may conclude that the concepts of attraction and repulsion which are central to the *Principia* and the *Opticks*, including that of "gravitational attraction which was probably the last one to appear, were primarily the offspring of alchemical active principles" (ANC, p. 224). He relies in particular on the draft *Conclusio* written for *Principia* I but never used, which argues that chemical phenomena are best understood in terms of short-range forces. Westfall maintains that this argument carries the main weight of the case for the concepts of attractive and repulsive force in general, and that the chemical processes described here were studied by Newton "in his own alchemical experiments". The use of the term 'alchemical' for Newton's experiments on chemical process is somewhat debatable. The experimental precision of his work in the laboratory and the careful language in which he reports his results (both detailed by Westfall) makes the term a risky one in this context, even though Newton undoubtedly was deep in alchemical reading at this time, occasionally employed alchemical imagery to describe what he was doing, and quite probably was using the experiments to test the empirical claims that underlay the alchemical allegories. It is notable that in the *Conclusio* he uses none of the traditional metaphor which characterizes his notes on alchemy. He simply describes a variety of chemical processes in prosaic and exact terms, and argues that they are best understood as due to "the forces of particles". The fact that his interest is alchemy seems to have waned at the time he moved to the Mint in 1696 may testify not so much to his fear that it would be taken amiss in his new more public setting as to his growing confidence in the explanatory potentialities of his new mechanics, even in regard to the hitherto hidden realm of chemical and vital process.

90. *Correspondence*, vol 2, p. 308. See D. T. Whiteside, *Geometry and the Dynamics of Motion*, vol. 6 of *The Mathematical Works of Isaac Newton*

(Cambridge, 1975), p. 13. In the introduction to this volume and in *BP*, Whiteside traces the gradual formulation of the problematic of the *Principia* and argues that the traditional account of the discovery of the inverse-square law in 1666 is a myth. It was in 1684, he maintains, that Newton finally moved away from planetary vortices to the formulation of the area law, which, as Proposition I, formed the starting point of the *Principia*. He argues that the construction of the *Principia* was primarily a mathematical accomplishment, but one with ontological and conceptual implications that later had to be faced. He notes Newton's "growing awareness of the verbal subleties and epistemological difficulties involved in creating a viable mathematical theory of forces and the accelerated motions these induce in bodies" (*op. cit.*, p. 93).

91. The *Conclusio* is transcribed and translated in *USP*, pp. 320–47. See p. 333.

92. Gerd Buchdahl suggests a useful methodological structure, inspired by Kant, for the analysis of Newton's problems with the intelligibility of his concept of force. See "Explanation and gravity".

93. Edleston, *Correspondence*, pp. 152, 154. The axiom is found in Book III, Prop. 5, cor. 1. In a draft from the 1690s, perhaps meant for the *Opticks*, Newton singled out the principle of mutual attraction as one which is derived as principles, in his view, *should* be derived, that is, by "drawing conclusions from experiments" (ULC, Add. 3970, f.479r).

94. *Principia*, p. 410. Further texts are cited in footnote 111 below.

95. Koyré is of the opinion, on the other hand, that the *only* way to understand Newton's response is to admit that, for him, "'attraction', all the pseudo-positivistic and agnostic talk notwithstanding, was a real force (though not a mechanical and perhaps not even a 'physical' one) by which bodies *really* acted upon each other (though not immediately through a void, but by means of an *immaterial* link or medium) and that this 'force' was somehow located in or connected with, these bodies and was also dependent on, or proportional to, their masses" (*NS*, pp. 280–81). To locate Newton's forces in bodies in so unqualified a way seems much too strong, however, as we shall argue below, particularly in view of what Newton has to say of "active principles". Incidentally, if bodies "really act" on one another, could the forces between them be "non-physical"?

96. Draft-material for the *Optice:* ULC, Add. 3970, f.620v. See also f.619r (McGuire, *FAN*, pp. 167–71).

97. Draft-material for 31(23), ULC, Add. 3970, f.620v.

98. Early draft for the Queries of 1717; ULC, Add. 3970, f.235r. This "electric spirit" will be discussed below in Chapter 4.

99. This will be discussed in more detail in Chapter 4 below.

100. *Principia*, Book III, Prop. 42, p. 542.

101. *Op. cit.*, p. 530.

102. Far from restricting the terms 'spirit' and 'spiritual' to the description of God or God's action, therefore, he more often followed the "common

acceptance" of the terms (ULC, Add. 3965, f.437v); passage quoted in Chapter 4 below. The Halls, adopting the narrower sense, note that the Newtonian notion of force "allows no more spiritual quality to forces than to the very existence of matter" (*USP*, p. 196). But in view of Newton's frequent contrasting of the inertness of matter with the quality of initiating motion that he takes to be characteristic of "spirit", the opposite claim could just as easily be made, especially if Newtonian forces are *also* described as "certainly immaterial" (*USP*, p. 197). Newton would not have been so confident about a third category, "immaterial", between "material" and "spiritual".

103. Cudworth in his *True Intellectual System of the Universe* (1678) argued against atomists and atheists (who find, he says, only "passive principles" in the universe) the need for active principles to explain the various sorts of change. Newton read and annotated this work and was undoubtedly influenced by it.

104. "However we cast about, we find almost no other reason for atheism than this notion of bodies having, as it were, a complete and absolute reality in themselves" (*De Gravitatione*, *USP*, p. 144). See also Q 28, pp. 369–70, and McGuire, *BV*, p. 228. Newton makes the First Mover responsible for the initial motion of the universe; he was never quite sure whether to make Him also a sustaining cause of (changes of) motion via "active principles". He also has Him intervene to ensure the stability of the planetary system.

105. It could also be described as "metaphysical" since it derived from a metaphysical doctrine of the nature of God and of God's relation with the world. But since this doctrine in turn depended (more perhaps for Newton than for Leibniz) upon the hints given about the nature of God in the Bible, it can properly be described as "theological" also.

106. *Leibniz–Clarke Correspondence*, p. 12; McGuire, *FAN*, p. 197.

107. The extent to which the world is governed by rational necessity (assumed to be in tension with divine freedom) had been a major issue in all discussions of scientific methodology from the thirteenth-century debate between Aristotelians and Augustinians onward. One can, for instance, see Descartes struggling with it in the *Discourse on Method*. See McMullin, "Empiricism and the Scientific Revolution," in *Art, Science and History in the Renaissance*, ed. C. Singleton (Baltimore, 1968), pp. 331–69, especially the final section: "The enigma of Newton".

Chapter 3: Is Gravity an Essential Property of Matter?

108. For references, see F. Cajori's appendix to his edition of the *Principia*, pp. 632–38.

109. *PL*, p. 298.

110. *Opticks*, p. cxxiii.

111. For references, see A. Koyré, "Gravity an essential property of matter?" NS, pp. 149–63, and note 6 of the Motte-Cajori translation.

112. Letter to Bentley, PL, pp. 302–3.

113. We shall return to this in Chapter 4.

114. McGuire suggests a further reason: "the essence of a thing must be a true or absolute property of it, which does not depend on the existence of other things. For Newton, both extension and *vis insita* fall into this category. Gravity, on the other hand, is a property which determines the spatial relations among bodies. It is in this sense not an essential attribute of matter, and Newton felt the need to explain it. The relational nature of gravity, then, was probably at the heart of his rejection of it as essential to matter" (CC, p. 189). This reason would hold good only if gravity be taken as actualized. To the extent that it is a disposition or capacity, it could be possessed even by a body existing in a universe otherwise empty.

115. See my introduction to *The Concept of Matter in Modern Philosophy*, section 7.

116. Book III, Prop. 41, p. 528.

117. ULC, Add. 3970, f.243v.

118. This ratio is invariant only if the earth be taken as a "given", as Newton appears to be assuming. If the distribution of other masses (sun, other planets, etc.) be taken into account, it is not invariant.

119. This tends to undercut McGuire's suggestion that Newton's reason for refusing to admit gravity as an essential quality of matter was its "relational nature". Newton, he argues, was committed to holding that "the essence of a thing must be a true or absolute property of it, which does not depend on the existence of other things. . . . Gravity . . . is a property which determines the spatial relations among bodies. It is in this sense not an essential attribute of matter" (CC, p. 189). But Newton never does specify that gravity is relational; on the contrary, he sometimes (as we have just seen) makes it intensity-invariant, implying that it is independent of changing external relations. To say that gravity "determines the spatial relations among bodies" is *not* to say that as a disposition it is itself dependent upon the existence of those bodies.

120. Prop. 7, Cor. I.

121. Newton did think of including this criterion here, but changed his mind. In his interleaved copy of the first edition in which the new sentence to be inserted in Corollary II first appeared, he added some more lines. Gravity, he says, is "proportional to the quantity of matter in each body and can neither be increased nor decreased in intensity. Hence by Hypothesis [Rule] III, it is a property of all bodies". He later canceled this addition; this perhaps reinforces our earlier suggestion (§1.2) that he was never quite easy about the intensity-invariance criterion, which, even though it is specified in the new Rule III as one of the two criteria for a universal quality, is never alluded to in the lengthy discussion and application of the Rule which follows its statement.

122. The retention of the intertransmutability argument in the later editions has a certain humor about it. It is, of course, no longer needed, since the introduction of Rule III made any further argument redundant. But what is more peculiar is that Hypothesis III, asserting universal intertransmutability (on which the argument directly depended), was dropped from these later editions and in its place was substituted Rule III (which asserts that certain qualities are invariant, i.e., do *not* lend themselves to any form of transmutation). In order to retain an argument which had been so thoroughly undermined, Newton hit on the ingenious resort of making it an independent *ad hominem* argument against those who, in his view, *did* accept the intertransmutability hypothesis (which he now by implication disavows). And so he adds a phrase attributing the earlier hypothesis to "Aristotle, Descartes and others". (Aristotle, for one, would never have admitted that quantity of matter is an invariant throughout all forms of change, nor that the only difference between bodies lies in "mere form of matter".) See footnote 15 above.

123. *Principia*, pp. 399–400.

124. Cor. II to Prop. 5, added in the second edition. See the KC edition, pp. 574ff., for details and for the draft passages translated here; see also *Principia*, p. 413.

125. *Principia*, p. 410.

126. *Op. cit.* p. 546.

127. *Principia*, p. 547.

128. Draft for projected addition to Query 31 (1716); ULC, Add. 3970, f.243v.

129. See §3.1 above, notes 109 and 112.

130. *Principia*, p. xxvi.

131. Letter from Cotes to Clarke, *Correspondence*, p. 151.

132. Koyré maintains that Cotes' initial use of the term 'essential' "shows how easy it was for a serious and well-informed student of Newton to misunderstand him": it was, he thinks, "an obvious error" (NS, pp. 159, 281). In view of the ambiguities in Newton's position, however, the error (if it was one) was certainly not obvious.

133. Book III, Prop. 6, Cor. II. Newton tried a different counterargument in an earlier piece (notes for the revision of the *Principia* in the 1690s). The explanation of gravity by means of particles lacking in gravity can be shown to entail that there cannot be density differences between the gravitating particles, and this in turn forces the suspension of the principle of universal transmutability (which at that period, as we have seen, Newton thought to be a basic philosophical principle). See McGuire, *TI*, pp. 72–73, for a transcription of the argument. (Change 'so as' to 'if they are not' in line 13 of his translation, and '*sic*' to '*ne*' in line 9 of the transcribed passage; ULC, Add. 3965.6, f.267r.)

134. *Leibniz–Clarke Correspondence*, p. 172.

135. In the *Opticks*, he several times speaks of bodies acting at a distance

"not only on the rays of light . . . but also upon one another for producing a great part of the phenomena of Nature" (Q 31[23]; see also Q 1). Though posed as a query, it is clear that he assumed such action to be possible. Nevertheless, he is still trying to cover himself, and to de-ontologize in an especially emphatic way his use of the term 'attraction': "What I call attraction may be performed by impulse or by some other means unknown to me. I use that word here to signify only in general any force by which bodies tend towards one another, whatsoever be the cause" (p. 376). He had made the same disclaimer earlier, in the *Principia*: "the reader is not to imagine that by [words such as 'attraction'] I anywhere take upon me to define the kind or the manner of any action, the causes or the physical reason thereof" (pp. 5–6); or again: "I here use the word 'attraction' in general for any endeavor whatever made by bodies to approach one another, whether that endeavor arise from the action of the bodies themselves as seeking each other or agitating each other as spirits emitted; or whether it arises from the action of the aether or the air or of any medium whatever" (p. 192).

136. *Leibniz–Clarke Correspondence*, p. 94; see also M. B. Hesse's paper "Action at a distance" in *The Concept of Matter in Modern Philosophy*.

137. General Scholium, p. 546. Newton did not consistently defend this view. In a late draft (c. 1716), he claimed that he had "nowhere denied that the cause of gravity is mechanical" (ULC, Add. 3968, f.125r).

138. "For gravity without a miracle may keep the planets in. And to understand this without knowing the cause of gravity is as good a progress in philosophy as to understand the frame of a clock and the dependencies of the wheels upon one another without knowing [that] the cause of the gravity of the weight which moves the machine is in the philosophy of clockwork". From a draft-letter (c. 1714) to the editor of the *Memoires de Trevoux*, reacting to Leibniz' views as expressed in the Hartsoeker–Leibniz correspondence in the *Memoires* for May 1712 (ULC, Add. 3968, f.257r). The letter was never sent.

139. *Leibniz–Clarke Correspondence*, p. 53.

140. *Ibid.*

141. *Principia*, p. xviii.

142. As the Halls, for example, tend to do. "He was a 'mechanical philosopher' in the seventeenth century sense" (*USP*, p. 75). See also p. 202 and their article "Newton's mechanical principles".

Chapter 4:   How Is Matter Moved?

143. ULC, Add. 3996, f.97r.

144. "No being can exist which is not in some way related to space. God is everywhere, created minds are somewhere, and body is in the space it occupies. Whatever is neither everywhere nor somewhere does not exist.

Hence it follows that space is an effect arising from the first existence of being". Hall and Hall, *USP*, pp. 103, 141.

145. *USP*, pp. 141–3. See also Westfall, *FNP*, p. 340.

146. The "Hypothesis explaining the properties of light" is reprinted in *PL*, pp. 178–99. See p. 180.

147. *PL*, p. 181.

148. *PL*, p. 184.

149. *PL*, p. 185.

150. It is worth noting that the explanation of the "inflection" of light (refraction, reflection), given in the Hypothesis, was still a mechanistic one, involving contact action operating over an aether density-gradient. This was a step beyond the single-collision theory of Descartes (who made the light particles collide with the particles of the reflecting/refracting body), but it still provided no grounds for the introduction of forces acting at a distance. It was not from optics that this idea ultimately derived but from gravitational mechanics. When Newton in the *Principia* finally shows how Snell's Law of refraction can be explained in terms of a constant force acting continuously on the light-particle over a short distance to alter its trajectory (Book I, Prop. 94), it is evident that the basic model is derived by analogy from the gravitational case, and not the reverse. See Bechler, *NMC*.

151. P. M. Rattansi argues that the alchemical/hermetic inheritance gave Newton's imagination a range it would otherwise have lacked, in his attempts to find a model for mechanical action. See *ER*, p. 165. See also Westfall, *FNP*, p. 466. This theme is much more fully discussed by several contributors to M. Righini Bonelli and W. Shea, eds., *Reason, Experiment and Mysticism in the Scientific Revolution*.

152. *USP*, p. 223. Westfall takes this paper to mark the watershed in the development of Newton's concept of force (*FNP*, p. 377).

153. McGuire (in *FAN*) gives a detailed review of the relevant published and unpublished material.

154. Addendum crossed out in a draft for Q 31(23); ULC, Add. 3970, f.252v. A draft for a "Query 17" (never used), written around the same time, suggests a quite different model, however: "Is there not something diffused through all space, in and through which bodies move without resistance and by means of which they act upon one another at a distance in harmonical properties of their distances?" (ULC, Add. 3970, f.234v.

155. McGuire suggests that for Newton at this time there are "two sorts of motion in nature", one characteristic of *vis inertiae* and its "'passive laws of motion" ("these are the laws of the *Principia*"), and the other "new" or "vital" motion due to active prinicples (*FAN*, p. 172). He goes on: "Newton conceived the possibility of determining the mode of action of laws not based directly on matter: laws of forces causing such motions as fermentation and putrescence, phenomena not reducible to the mere translation of matter or to mechanical impulse. Since forces causing these motions exist separately from matter, these laws should also be independent of material phenomena"

(p. 173). He concludes that Newton was moving to the view that a natural philosophy could be "based more directly on the phenomena of force" than on matter-properties; it would be a different "type of physical knowledge" than the one embodied in "the *Principia*, with its rigorous treatment of the motions of gross bodies" (p. 173; see also p. 161).

156. Draft for Q 31(23); ULC, Add. 3970, f.620v. David Gregory writes in 1705, after a conversation with Newton about the Queries which he was then preparing for the *Optice*: "His doubt was whether he should put the last Quaere thus: What the space that is empty of body is filled with? The plain truth is that he believes God to be omnipresent in the literal sense; and that as we are sensible of objects when their images are brought home within the brain, so God must be sensible of every thing, being intimately present with every thing: for he supposes that as God is present in space where there is no body, he is present in space where a body is also present. But if this way of proposing this his notion be too bold, he thinks of doing it thus: What cause did the Ancients assign of gravity? He believes that they reckoned God the cause of it, nothing else, that is, no body being the cause, since every body is heavy". W. Hiscock, *David Gregory, Newton and Their Circle* (Oxford, 1937), p. 29; quoted by Thackray, *AP*, p. 27. McGuire and Rattansi marshal numerous passages from the draft-material to indicate how characteristic it was of Newton during this period to make God's direct action the source of change in the natural order (*NPP*, §2). See also Kubrin.

157. The introduction at this time by Newton of what Bechler (*NLF*) calls "inverse-mass forces" such as magnetism, where "attraction is strongest in the smallest bodies in proportion to their bulk", does not affect this claim. First, Newton is not talking here of the *effects* of force but of the *source* of force: the smaller the body, the greater the ratio of the force it can produce (to move another body) in proportion to its own bulk. Second, he is talking not of forces but of force/bulk ratios, a very different matter, as we shall see.

158. *Opticks*, p. 401.

159. "These forces may be reckoned among the laws of motion and referred to an active principle". In a draft for Q 31(23); ULC, Add. 3970, f.620v.

160. In *FAN*, McGuire suggested that for Newton at this time there were "four fundamental types of entities existing in void space", matter, motions, forces, and active principles (p. 172). In particular, Newton "consistently held that forces were caused by agents which were characterized by the general rubric of active principles" (p. 186). He associates this with the doctrine of the "chain of being" (see §1.3 above); "relative to matter, force was non-material, but relative to its cause it was material". This is a neat schema, but it is risky to suggest that Newton consistently employed it. At best, his terminology in this regard was ambiguous and shifting.

161. Draft conclusion for the *Opticks* from around 1692. ULC, Add. 3970, f.336r (Bechler, *NLF*, p. 197).

162. One might be tempted to read Newton's frequent observation

around this time, that small magnets attract more in proportion to their size than large ones do (references in Bechler, *NLF*), as implying an *inverse* correlation between mass and magnetic force. But this is not what Newton said: a larger magnet will always (other things being equal) attract more strongly than a small one will. It is the *ratio* of the force produced by the magnet to the size of the magnet that increases with decreasing size of the magnetic body. Were one to adopt the other reading, Newton would be immediately saddled with an inconsistency. The passage just quoted (footnote 161) indicates that Newton accepted an additivity law for magnetic force; it appears unlikely that he would have contradicted this by the claim that a single particle of the magnet could exert a greater force than the whole magnet.

163. *Principia*, Book III, Prop. 6, Cor. V.

164. Q 31(23), *Opticks*, pp. 401–2.

165. ULC, Add. 3970, f.292r (Bechler, *NLF*, p. 201).

166. Q 30, *Opticks*, p. 374.

167. He later realized that this was an error, but only after the *Optice* had appeared. The force is proportional to the *square* of the velocity, not the velocity. See Bechler, *NLF*, p. 220.

168. F.292r. This passage was incorporated into Query 22 of the *Optice*, with some small changes. It was suppressed in the *Opticks* of 1717.

169. Bechler argues (*NLF*, p. 199) that they would be incompatible with the dynamic theory of the *Principia*. But Newton explicitly recognized the possibility in the *Principia* of non-mass-dependent forces (such as magnetism); the strict proportionality of force and mass in the *attracting* body could well be peculiar to gravitation. The already identified attractions (gravitational, magnetic, electric) "show the tenor and course of Nature and make it not improbable but that there may be more attractive powers than these" (Q 31); these presumably follow different laws, while nevertheless obeying the general force/inertial-mass relation defined in the three Axioms of the *Principia*.

170. Book III, Prop. 6, Cor. 4.

171. ULC, Add. 3970, f.336r.

172. Book I, Prop. 85.

173. Bechler suggests (*NLF*, pp. 198–99) that such short-range forces would involve a "screening effect" whereby the outer particles of a body would "screen" or "filter off" the forces emanating from the interior particles. This could imply a breakdown in the proportionality between the force and the mass of the complete body. Since the force would depend only on the outer particles, the rest serving as "inert ballast", he infers that the force/mass ratio would then be *inversely* related to the mass of the body; this, he thinks, may well be what Newton had in mind. First, however, Newton *cannot* allow such "screening" to occur. It is essential to the entire theory of force in the *Principia* that particles should be entirely transparent to the gravitational forces propagated through them. Second, it is unnecessary; the

interior particles will be ineffective, not because they are inert or because their forces have been screened off but because they are out of range; that is, their forces drop off so sharply as to be imperceptible outside the body. Third, the familiar force/total-mass proportionality *will* break down in these cases, but there is no reason to suppose that this ratio will be inversely related to the mass instead. Fourth, there is no evidence that Newton had anything like a screening effect in mind.

174. This would happen with an inverse-square law only if both test-particles are point-masses.

175. Book I, Sections XII and XIII; see especially Prop. 86 and the Scholium after Prop. 78.

176. It does, but it would have been very difficult for Newton to work this out by means of the geometrical formalism he was using.

177. Bechler (NMC) notes that he does not mention it after 1675.

178. ULC, Cdd. 3970, f.476. Bechler notes (*NLF*, p. 194) that after the appearance of the *Principia*, Newton asked Flamsteed (without explaining why) whether the light from Jupiter's satellites showed any change of color during their eclipses. Presumably, therefore, he was still entertaining the velocity model as a possibility at this time. But Flamsteed's negative answer may have finally decided him to try a different model.

179. ULC, Add. 3970, f.291r (Bechler, *NLF*, p. 194). This idea had, in fact, occurred to him much earlier. In the draft conclusion for *Principia* I he had written (and then scored out): "But that all the phenomena of colours arise from the different sizes of the particles can be proved by certain experiments", *USP*, p. 336.

180. See also drafts at f.292r and 259r.

181. Draft for *Opticks* III. ULC, Add. 3970, f.621r (Bechler, *NLF*, p. 207).

182. *Opticks*, p. 372.

183. This was the kernel of truth that permitted Newton to retain this argument throughout the re-editions and new drafts. The model was *not* an "aberration" in terms of Newton's earlier mechanics (as Bechler urges, *NLF*, p. 190), and so no special explanation (of the sort Bechler proposes) is needed for its retention.

184. This is Bechler's expression, and in *NLF* and *NMC* he argues at length for the propriety of its use in this context.

185. Bechler regards the "inverse-mass" model as a "subtle break with the doctrine of the *Principia*"; he asserts that there can be "no doubt" that in the *Opticks* Newton was offering something quite "revolutionary", namely, that the proportionality of weight and mass "so vehemently argued for" in the *Principia* was "after all only approximate" (*NFL*, pp. 212–3). If this were correct, it would involve an extraordinarily far-reaching re-evaluation of Newton's achievement in mechanics. Our analysis above suggests that it is not correct.

186. See A. R. Hall and M. B. Hall, "Newton's electric spirit", McGuire, *FAN*, notes 66, 67, and p. 185–86.

187. Draft-material for *Opticks* III. ULC, Add. 3970, f.235r.

188. A Latin draft from this period, ULC, Add. 3970, f.604r.

189. McGuire, *FAN*, pp. 175–78. Among other things, he suggests that all bodies "may abound with a very subtle, but active potent electric spirit by which light is emitted, refracted and reflected" (draft for Q 25; ULC, Add. 3970, f.235r).

190. The phrase 'electric and elastic' was added by Motte in translation but is faithful to the intent of the passage; *Principia*, p. 547.

191. See H. Guerlac, "Newton's optical aether"; L. Rosenfeld, "Newton's views on aether and gravitation"; E. J. Aiton, "Newton's aether-stream hypothesis"; P. Heimann and J. E. McGuire, *NF*.

192. Draft for a second edition of the *Principia*, dating from the 1690s; Library of the Royal Society, Gregory *MS*, 247, f.14a (McGuire, *FAN*, p. 163).

193. ULC, Add. 3968, f.257v. He suggests this only as the sort of possibility that Leibniz should not reject on the grounds of its being occult or a fiction or a miracle. But in *Principia* II he inserted a passage which refers to the earth as "floating in the non-resisting aether" (Book I, Cor. VI to Axioms, Scholium, p. 26).

194. See the drafts cited by McGuire, *BV*, pp. 217–22.

195. It may be the case that the notion of "electric spirit" he put forward to explain the high-intensity cohesive forces exhibited in these experiments suggested the rather different notion of an optical and gravitational aether, though the passage cited by Guerlac (*op. cit.*, p. 48) to show this is not persuasive. (It refers to an "aether or spirit" which operates close to bodies, i.e., the electric spirit; this is not the all-pervasive medium of *Opticks* III.) Whether it was *suggested* by the earlier work or not, the Hauksbee experiments could hardly be regarded as evidence in support of the 1717 aether.

196. Guerlac, *op. cit.*

197. ULC, Add. 3965, f.641r.

198. A revision of Q 31; the first and last sentences occur substantially in the published text (p. 394). From the draft *Observations* written for the 1717 *Opticks* (ULC, Add. 3970, f.622r).

199. In the General Scholium there is no hint of the electric spirit's itself being particulate: "by [its] force and action, the particles *of bodies* attract one another at near distances". In Query 22, however, he speaks of "how an electrick body can by friction emit an exhalation so rare and subtle and yet so potent as by its emission to cause no sensible diminution of [its] weight". Here the "exhalation" is represented in corporeal terms with some suggestion of its being particulate ('subtle' was most often used to emphasize the small size of constituent particles). Newton quite evidently moved easily from one to the other model, which is not too surprising since he regarded them both as quite speculative.

200. Thus indicating that he is thinking of density not in terms of number of (possibly immaterial) particles per unit volume but as a measure of *vis inertiae*. In Definition I of the *Principia*, when defining quantity of matter in

terms of density and volume, he had explicitly excluded aether: "I have no regard in this place to a medium, if any such there is, that freely pervades the interstices of bodies".

201. Q 21, p. 351. One obvious inadequacy is that this would at best account for the motion of only *one* of the bodies, whereas the principle of mutual attraction demands an explanation for the motions of both.

202. Intended for a new Part II in Book III of *Opticks* III; ULC, Add. 3970, f.623r.

203. One of the draft definitions of "body" for a section that was never added to the third edition of the *Principia*, dated tentatively by McGuire at 1716 (BV, pp. 218–19); ULC, Add. 3965, f.437v.

204. As illustrated in much recent writing on eighteenth-century science; see, for example, Arnold Thackray, *Atoms and Powers: An Essay on Newtonian Matter-Theory and the Development of Chemistry*; Robert Schofield, *Mechanism and Materialism: British Natural Philosophy in an Age of Reason*; Heimann and McGuire, "Newtonian forces . . ."; P. Heimann, "Nature is a perpetual worker."

205. This is the commonest view: that Newton regarded action at a distance as unintelligible, and bent every effort to eliminate it from his system. (A modified form of this would have him motivated not so much by his own conviction regarding action at a distance as the general disfavor with which it was regarded.) See, for instance, G. Buchdahl, *HS*.

206. The view that Newton's basic premise was the passivity of matter is suggested by A. Gabbey (*op. cit.*, p. 47), for example. "The passivity of body can be taken as the prior empirical datum", since it is "empirically demonstrated"; it is prior to any knowledge we have of the source of agency. McGuire also stresses this premise, expecially in *FAN*.

207. *PL*, p. 302.

208. *Op. cit.* pp. 375–76.

209. McGuire attributes a more definite and independent reality to Newton's forces (in the period up to 1706, at least) than do most other writers. Yet he notes: "No straightforward answer can be given as to whether Newton's forces are material or immaterial. They are manifestly different from both matter and spiritual substance, and thus they seem to occupy a twilight zone between the corporeal and the incorporeal. Relative to ponderable bodies they are incorporeal; but relative to the higher realms of the spiritual they are corporeal" (*FAN*, p. 185). The "twilight" here is due even more to the haze surrounding these three pairs of contrast terms (in Newton's usage of them) than it is to the uncertainty regarding the status of force. We must be cautious about supposing that Newton thought of forces as entities or agencies in their own right, and that the problem lies mainly in finding the proper categories in characterizing them. See footnotes 155, 160 above.

210. L. Laudan, in a comment accompanying Buchdahl's *HS* argues for a stronger thesis, namely, that for Newton: "action at a distance is intelligible and defensible *per se*, without falling back on arguments about design, the

harmony and tenor of nature, and the like" (p. 238). He correctly interprets the Bentley letter (quoted above) as implying that the bodies *can* act upon one another at a distance, provided that it be "through the mediation of something else, which is not material, . . . by and through which action and force may be conveyed from one to another". But he wrongly poses this as an *alternative* reading to one which would have Newton claim the inconceivability of action at a distance. Newton is saying *both* things. The operation of the "something else which is not material" is precisely *not* action at a distance. In fact, in this text (unlike others) Newton is asserting that even immaterial agencies work by "conveying" force and action from one place to another; that is, they do not work "at a distance". The Bentley letter in no way, therefore, supports Laudan's thesis of the total acceptability to Newton of action at a distance; it is still the counterargument to that view it has always been assumed to be. But Laudan is right in saying that Newton's later receptivity toward the sort of active principles that *do* seem to operate "at a distance" (in some sense, at least) indicates that the core of his position can hardly be taken to be opposition to the idea of action at a distance *per se*, only to that of *bodily* action at a distance.

211. Such texts come mainly from the decade 1695–1705: "As all the great motions of the world depend upon a certain kind of force (which in this earth we call gravity) whereby great bodies attract one another at great distances; so all the little motions in the world depend upon certain kinds of forces whereby minute bodies attract or dispel one another at little distances" (material from the 1690s for a projected Book IV of *Opticks*; ULC, Add. 3970, f. 338r). There is no hesitation in this draft about attributing activity to matter directly. Did he have some sort of divine "back-up" action in mind? After 1706, when the "spirit" metaphor begins to dominate, one rarely finds him attributing force directly to matter. Rather, he will seek the "agents" or "mediums" in Nature which are able to make "the particles of bodies attract one another" (ref. at footnote 193).

212. This is the suggestion of L. Laudan (*op. cit.*, pp. 237–38). He rejects, as we have seen, the view that Newton was guided by P1 (footnote 210) and does not discuss P2.

213. McGuire says of the 1717 aether that it is "difficult to suppose that Newton took it seriously. It was subject to obvious conceptual inconsistencies; it was a flagrant example of the sort of intermediary entity which Newton had always tended to reject; and more significantly it repudiated his basic metaphysics of God in an empty universe... [it] remains the most puzzling of Newton's attempts at the problem of the cause of force" (*FAN*, p. 187). Laudan, on the contrary, argues that "Newton found the ether to be a concept of high explanatory value and relied on it increasingly as he found more and more kinds of phenomena which it seemed to explain", notably the phenomena of radiant heat, damping of pendulum motion, and optical transmission. The differences between these points of view can be diminished somewhat (not entirely!) by noting that McGuire is thinking of the

aether mainly in the context of gravitational explanation, whereas Laudan is focusing on all the *other* contexts.

214. *PL*, p. 303.

215. *Principia*, Book I, Prop. 69; Book II, Prop. 24; Book III, Prop. 6. Of course, it was the impressive *wholeness* of Newton's account, rather than its ability to deal with this context or that, which won the day for it.

Chapter 5:   Epilogue: Matter and Activity in Later Natural Philosophy

216. See my introduction to *The Concept of Matter in Modern Philosophy*.

217. For helpful discussions of Greene's views, see Heimann and McGuire, *NF*, pp. 254–61; Schofield, *MM*, pp. 117–21. Greene formulated his position more fully in a later work, *The Principles of the Philosophy of the Expansive and Contractive Forces* (1727).

218. *Principles of . . . Forces*, p. 286 (quoted in *NF*).

219. *Ibid.*, p. 659.

220. *Ibid.*, p. 62.

221. Traced in some detail by Thackray in *MN*.

222. The *Opticks* of 1704 devotes some space to this argument (Book II, Part III).

223. A note added to the 1715 edition of his *Introductio ad Veram Physicam*, quoted by Thackray, *MN*, p. 41.

224. *Leibniz–Clarke Correspondence*, p. 11.

225. *Letters Concerning the English Nation*, p. 147; quoted by Thackray, *MN*.

226. *Principles of Human Knowledge* (1710), par. 25.

227. *Ibid.*, par. 103–5.

228. *Dissertation on the Aether of Sir Isaac Newton*, p. 122; see Thackray, *AP*, p. 138.

229. See Heimann, *NPW*.

230. *Op. cit.*, trans. J. M. Child (Cambridge, Mass., 1966), p. 183.

231. *Op. cit.*, p. 21.

232. *Op. cit.*, p. 183.

233. *Op. cit.*, r. 187.

234. *Op. cit.*, p. 188.

235. *The Theological and Miscellaneous Works of Joseph Priestley*, ed. J. T. Rutt (London, 1817–31), vol. 3, p. 220. Cited in Heimann and McGuire, *NF*, p. 270. See also Schofield, *MM*, pp. 261–73.

236. *Op. cit.*, p. 230.

237. *Op. cit.*, p. 219.

238. See Heimann and McGuire, *NF*, Schofield, *MM*, and Thackray, *AP*, for a discussion of such scientists as Hutton, Stewart, Playfair, Laplace.

239. *Op. cit.*, trans. J. Ellington (Indianapolis, 1970), p. 43. Chapter 2 ("Metaphysical Foundations of Dynamics") is entirely devoted to the relations of matter and force.

240. *Op. cit.*, p. 90.

241. For a detailed discussion of this text, see Mary Hesse, *Forces and Fields*, Chapter 7.

242. *Metaphysical Foundations*, Prop. 7 of Chapter 2, p. 61.

243. *Op. cit.*, pp. 64–65.

244. *Op. cit.*, p. 66.

245. *Op. cit.*, p. 78; see also p. 62.

246. See Mary Hesse, "Action at a distance", in McMullin, ed., *The Concept of Matter in Modern Philosophy*, and her *Forces and Fields*, Chapter 8.

247. See E. McMullin, "The fertility of theory".

References

[The following works are cited in the footnotes. Abbreviations are specified in some cases. For a fuller bibliography and a useful general account of Newton's work, see I. B. Cohen, "Newton," *Dictionary of Scientific Biography* (1974). References to the draft-material catalogued as "Additional mss." in the University Library of Cambridge are given as "ULC, Add.," with the folio and page numbers. The manuscripts are on the whole quite legible and so there is rarely any ambiguity in the transcription. Words scored out by Newton are not transcribed; words inserted by him above the line are, of course, included. I have modernized spellings and punctuation. Though much has been done by I. Bernard Cohen, A. R. and M. B. Hall, J. E. McGuire, and others, much draft-material still awaits publication.]

AITON, E. J. "Newton's aether-stream hypothesis", *Ann. Science*, 25, 1969, 255–60.

ALEXANDER, H. G. ed. *The Leibniz–Clarke Correspondence.* Manchester, 1956.

BECHLER, Z. "Newton's law of forces which are inversely as the mass...", *Centaurus*, 18, 1974, 184–222. Abbrev. *NLF*.

———. "Newton's models of color dispersion", *Arch. Hist. Exact Sciences*, 11, 1974, 1–37. Abbrev. *NMC*.

BUCHDAHL, G. "History of science and criteria of choice", *Minn. Studies Phil. Science*, 5, 1970, 204–30. Abbrev. *HS*.

———. "Explanation and gravity", in *Changing Perspectives in the History of Science*, ed. M. Teich and R. Young. London, 1973, 167–203.

COHEN, I. B. *Franklin and Newton*. Cambridge, Mass., 1956. Abbrev. *FN*.

——; ed. *Isaac Newton's Papers and Letters on Natural Philosophy*. Cambridge, 1958. Abbrev. *PL*.

——. "Hypotheses in Newton's philosophy", *Physics*, 8, 1966, 163–84. Abbrev. *HNP*.

——. "Newton's Second Law and the concept of force in the *Principia*", in *Annus Mirabilis* (see Palter), pp. 143–85. Abbrev. *NSL*.

DOBBS, B. J. T. *The Foundations of Newton's Alchemy or the Hunting of the Greene Lyon*. Cambridge, 1975.

EARMAN, J. and M. FRIEDMAN. "The meaning and status of Newton's Law of Inertia . . .", *Philos. Science*, 40, 1973, 329–59.

EDLESTON, J. *Correspondence of Sir Isaac Newton and Professor Cotes*. London, 1850.

ELLIS, B. "The origin and nature of Newton's laws of motion", in *Beyond the Edge of Certainty*, ed. R. Colodny. Englewood Cliffs, N.J., 1965, pp. 29–68.

GUERLAC, H. *Newton et Epicure*. Paris, 1963.

——. "Newton's optical aether", *Proc. Royal Soc.*, NR, 22, 1967, 45–57.

HALL, A. R., and M. B. HALL. *Unpublished Scientific Papers of Isaac Newton*. Cambridge, 1962. Abbrev. *USP*.

——. "Newton's mechanical principles", *Journal Hist. Ideas*, 20, 1959, 167–78.

——. "Newton's electric spirit—four oddities", *Isis*, 20, 1959, 473–76.

HAWES, J. "Newton's two electricities", *Ann. Science*, 27, 1971, 95–103.

HEIMANN, P. M. "Nature is a perpetual worker: Newton's aether and 18th century natural philosophy", *Ambix*, 20, 1973, 1–25. Abbrev. *NPW*.

—— and J. E. MCGUIRE. "Newtonian forces and Lockean powers: Concepts of matter in 18th century thought", *Hist. Studies Phys. Sciences*, 3, 1971, 233–306. Abbrev. *NF*.

HERIVEL, J. *The Background of Newton's Principia*. Oxford. 1965.

HESSE, M. *Forces and Fields*. Edinburgh, 1961.

JAMMER, M. *Concepts of Mass*. Cambridge, Mass., 1961.

KOYRÉ, A. *Newtonian Studies*. Chicago, 1965. Abbrev. *NS*.

—— and I. B. COHEN. "Newton and the Leibniz–Clarke correspondence", *Archiv. Intern. Histoire Sciences*, 15, 1962, 63–126.

KUBRIN, D. "Newton and the cyclical cosmos: Providence and the mechanical philosophy", *Journal Hist. Ideas*, 28, 1967, 325–46.

MANDELBAUM, M. *Philosophy, Science and Sense-Perception*. Baltimore, 1964.

McGUIRE, J. E. "Body and void and Newton's *De Mundi Systemate*: Some new sources", *Arch. Hist. Exact Sciences*, 3, 1967, 206–48. Abbrev. *BV*.

——. "Transmutation and immutability: Newton's doctrine of physical qualities", *Ambix*, 14, 1967, 69–95. Abbrev. *TI*.

——. "Force, active principles and Newton's invisible realm", *Ambix*, 15, 1968, 154–208. Abbrev. *FAN*.

——. "The origin of Newton's doctrine of essential qualities", *Centaurus*, 12, 1968, 233–60. Abbrev. *EQ*.

——. "Atoms and the Analogy of Nature: Newton's Third Rule of Philosophizing", *Studies Hist. Phil. Science*, 1, 1970, 3–58. Abbrev. *AAN*.

——. Comment on Cohen, *NSL*, in *Annus Mirabilis* (see Palter), pp. 186–91. Abbrev. *CC*.

—— and P. M. RATTANSI. "Newton and the pipes of Pan". *Proc. Royal Soc.*, *NR*, 21, 1966, 108–43. Abbrev. *NPP*.

McMULLIN, E., ed. *The Concept of Matter*. Notre Dame, 1963.

——. "The fertility of theory and the unit for appraisal in science", *Boston Studies in the Philosophy of Science*, 39, 1976, 395–432.

——. ed. *The Concept of Matter in Modern Philosophy*. Notre Dame, 1978.

NEWTON, Isaac. *Philosophiae Naturalis Principia Mathematica*. Third edition (1726). Trans. A. Motte and rev. F. Cajori. Berkeley, 1934. Critical edition ed. A. Koyré and I. B. Cohen. Cambridge, 1972. Abbrev. *KC*.

——. *Opticks*. Fourth edition, reprinted New York, 1952.

PALTER, R., ed. *The Annus Mirabilis of Sir Isaac Newton, 1666–1966*. Cambridge, Mass., 1970.

PAMPUSCH, Anita, C. S. J. "Isaac Newton's notion of scientific explanation". Ph.D. dissertation, Ann Arbor Microfilms, 1971.

RATTANSI, P. M. "Some evaluations of reason in sixteenth- and seventeenth-century natural philosophy", in *Changing Perspectives in the History of Science*, ed. M. Teich and R. Young. London, 1973, pp. 148–66. Abbrev. *ER*.

RIGHINI BONELLI, M., and W. SHEA, eds. *Reason, Experiment and Mysticism in the Scientific Revolution*. New York, 1975.

Rosenfeld, L. "Newton's views on aether and gravitation", *Arch. Hist. Exact Sciences*, 6, 1969, 29–37. Abbrev. NV.

Schofield, R. *Mechanism and Materialism: British Natural Philosophy in an Age of Reason*. Princeton, 1970.

Thackray, A. "Matter in a nutshell: Newton's *Opticks* and 18th-century chemistry", *Ambix*, 15, 1968, 29–53. Abbrev. *MN*.

———. *Atoms and Powers: An Essay on Newtonian Matter—Theory and the Development of Chemistry*. Cambridge, Mass., 1970. Abbrev. *AP*.

Turnbull, H. W., et al., ed. *The Correspondence of Isaac Newton*. Cambridge, 1959.

Westfall, R. W. *Force in Newton's Physics*. London, 1971, Abbrev. *FNP*.

———. "Alchemy in Newton's career", in *Reason, Experiment and Mysticism*, ed. Righini Bonelli and Shea, 189–232. Abbrev. ANC.

Whiteside, D. T., ed. *The Mathematical Works of Isaac Newton*. Cambridge, 1967.

———. "Before the *Principia*: The maturing of Newton's thoughts on dynamical astronomy, 1664–1684", *Journal Hist. Astron.*, 1, 1970, 5–19. Abbrev. *BP*.

Index